A Student Compan

Genes VI

A Student Companion and Workbook for

Genes VI

BENJAMIN LEWIN

By

Martin G. Klotz

Assistant Professor, Department of Biology, University of Colorado, Denver, USA

Paul G. Siliciano

Assistant Professor, Department of Biochemistry and Institute of Human Genetics, University of Minnesota, Minneapolis, USA

Oxford New York Tokyo
OXFORD UNIVERSITY PRESS
1998

Oxford University Press, Great Clarendon Street, Oxford OX2 6DP

Oxford New York
Athens Auckland Bangkok Boġota Bombay
Buenos Aires Calcutta Cape Town Dar es Salaam
Delhi Florence Hong Kong Istanbul Karachi
Kuala Lumpur Madras Madrid Melbourne
Mexico City Nairobi Paris Singapore
Taipei Tokyo Toronto Warsaw

and associated companies in
Berlin Ibadan

Oxford is a registered trade mark of Oxford University Press

Published in the United States by
Oxford University Press Inc., New York

A catalogue record for this book is available from the British Library

Library of Congress Cataloging in Publication Data
(Data available)

ISBN 0 19 857814 8

Typeset by the Author
Printed in Great Britain by
The Bath Press, Avon

Contents

Part 6: Eukaryotic gene expression 141

Part 7: Cell growth, cancer, and development 182

Answers

How to use this book

The Companion to Genes VI is intended for use by a variety of students, from middle- and upper-level undergraduates taking a first course in molecular biology to graduate students in advanced courses. Hence, we hope that this workbook will challenge students of all levels. The questions are purposely not arranged in order of difficulty and we recommend that students at every level try all the questions; they may be easier than they look!

This book is organized into Parts, following the layout of Genes VI. Each Part begins with a discussion of the major themes to this part, hence, they are among the most important sections of this book. All too often, students studying molecular biology concentrate on remembering that TFIID and TFIIIB contain TBP. In doing so, they fail to see the unifying principles of molecular biology. The Themes sections highlight these principles and tie them together throughout the book. When you start a new Part, read the Themes section, and go back to reread the previous Themes sections.

The Key words section lists important terms from each chapter: you should know the significance of all these words. In addition, some of them will be useful in completing the Summary worksheets found at the end of each chapter.

Your knowledge and understanding will be tested by the Multiple choice and Concept questions sections. For these questions, full answers are provided. For the multiple choice questions, try also to discover why each incorrect answer is wrong.

For the Problem solving exercises, few direct answers are provided. In some cases, hints are given; in others, not. These exercises are intended to make you think and to use the knowledge you have gained. For many of the problems, more than one correct answer is possible. For some others, no answers are known. This reflects the way research scientists approach these same questions: absolute answers may not be available, we have to use our knowledge and experience to gain as much insight as we can.

The Summary worksheet provides a synopsis of the important information in each chapter. These are the easiest types of questions in the Companion to Genes VI. You should be able to complete the narrative by filling in the blanks.

Note that because the subjects in Part 7 are beyond the scope of most courses in molecular biology, this material is not covered in the same depth as the rest of the book.

P.G.S. thanks the students in MdBc 8002, who served as guinea pigs (sometimes without their knowledge!), Patricia Hilleren, Mark Murphy, and Keiko Sakai for comments, and Howard Towle, Vivian Bardwell, David Zarkower, and Harry Orr for discussions. I wish to dedicate this writing to Dee, Robert, and Marie in thanks for their patience and support. Finally, special thanks to my mother, Ann M. Siliciano, who proofread every question, and who can answer every one of them too!

How to use this book

M.G.K. is thankful to the students in his 96 and 97 classes at UCD who had to cope with many of the questions in exams and discussions. Special thanks to Renea Hartwick, Ellen Levy, Jeanette Norton, Gregory Podgorski, and Jon Takemoto for comments and discussions. Thanks for the support from my family, friends, colleagues and students who have seen me through hard drive failures, a broken and a sprained wrist and a torn meniscus during the writing for this book. I wish to dedicate this writing adventure to my parents, Ruth Cecilie and Prof. Gerhard Franz Klotz on the occasion of their 70th birthdays.

It has been a pleasure working with the Oxford University Press staff.

Introduction

Theme

Before you dive into the details of DNA structure and function, their modification and the regulation of this modification, the two introductory chapters aim to remind you of the general picture of the cell from an architectural and biochemical point of view. You will notice a lot of bold type-faced words, a feature that continues throughout the entire textbook. Be advised that it is helpful for your studies to make these selected words part of your applicable vocabulary. Like in every other field, molecular geneticists have to speak the same language in order to communicate successfully. It is no contribution to the field or service to the learning student if an ambiguous undefined terminology is used. The study of the structure and function of genes requires knowledge of at least general biology and chemistry, but you are advised to have advanced knowledge in various aspects of cell biology, biochemistry and general genetics. Also, be forewarned that not everything in the field of molecular genetics has been 'worked out', hence you will encounter at various points rhetorical as well as real questions which reflects the fact that current knowledge is more like 'the last frontier' in a process of active exploration and enquiry. What you will also learn is that any kind of hypothesis is a valid hypothesis as long as it is not in conflict with established theories or as long as contradictory evidence has not been obtained repeatedly.

Key terms

Use the following key terms as a guide for your study and the composition of the summary at the end of each chapter.

Chapter 1

❑ catalysis	❑ glycosylation	❑ protein
❑ catalytic activity	❑ hydrophobic effect	❑ secondary structure
❑ conformation	❑ lipid	❑ signal transduction
❑ covalent bond	❑ polymerization	❑ structure
❑ enzyme	❑ polypeptide	❑ *trans*-acting
❑ genetic information	❑ polysaccharide	

Chapter 2

❑ assembly	❑ diploid	❑ meiosis
❑ cell growth	❑ envelope	❑ mitosis
❑ cellular structures	❑ eukaryote	❑ plasma membrane
❑ centromere	❑ expression	❑ processing
❑ cytoskeleton	❑ Golgi	❑ prokaryote

Chapter 1: Proteins

Multiple-choice questions

(1) **Which of the following could you substitute for asparagine without changing the overall charge of the polypeptide at a given pH value**

    (a) aspartate   (b) lysine   (c) histidine

    (d) tyrosine   (e) serine

(2) **Which of the following bonds can be considered non-covalent?**

    (a) α 1,4-glycosidic bonds

    (b) β 1,4-glycosidic bonds

    (c) disulfide bonds

    (d) hydrogen bonds

    (e) peptide bonds

(3) **Which of the following are recognized cellular macromolecules**

    (a) organic acids   (b) sulfuric acid   (c) nucleic acids

    (d) fatty acids   (e) amino acids

(4) **A phosphatase removes a phosphate moiety enzymically, while a kinase adds a phosphate group to another molecule. An enzyme that can carry out both functions is called a phosphorylase.**

    (a) true   (b) false

(5) **In proteins, the term 'secondary structure' refers to:**

    (a) the linear sequence of amino acids.

    (b) the three-dimensional organization of all atoms in the polypeptide chain.

    (c) the folding and intra-polypeptide interactions of the polypeptide chain (the path of its backbone).

    (d) the interaction between polypeptide chains that constitute a multimeric protein.

    (e) the relative content of charged versus uncharged amino acids in a protein.

(6) **The -SH functional group of cysteine residues can:**

    (a) contribute to the formation of peptide bonds at polypeptide branch points, hence, cysteine is a diamino acid.

    (b) interact with *N*-acetyl-glucose amine thereby initiating S-linked glycosylation.

    (c) engage in the formation of ionic bonds with salts.

    (d) engage in the formation of disulfide bridges thereby connecting different polypeptide chains.

    (e) engage in the formation of disulfide bridges thereby connecting different part of one polypeptide chain.

(7) **Glycoproteins are:**
   (a) enzymes involved in the catabolic processing of sugars.
   (b) proteins that are glycosylated with a side chain of fatty acids.
   (c) enzymes involved in the formation of glycosidic bonds.
   (d) proteins involved in protein glycosylation.
   (e) proteins that are glycosylated with a side chain of (N-linked or O-linked) oligosaccharides.

(8) **The spatial structure of all polar compounds in a cell is determined by the hydrophobic effect exerted by water molecules.**
   (a) true   (b) false

(9) **A ligand is:**
   (a) the active site in an enzyme that binds covalently to the substrate.
   (b) the active site in an enzyme that binds noncovalently to the substrate.
   (c) a general term for a small molecule that interacts with a protein in a covalent manner.
   (d) a general term for a small molecule that interacts with a protein in a noncovalent manner.
   (e) a globular protein lacking an enzymic function.

(10) **Once a polypeptide chain has been synthesized by the ribosome,**
   (a) it always is concomitantly folded into its functional structure.
   (b) it usually interacts with molecular chaperones that direct the folding process.
   (c) it usually interacts with molecular chaperones that direct the folding process or maintain the polypeptide in a conformation that allows for translocation.
   (d) it leaves the nucleus to interact with molecular chaperones that direct the folding process.
   (e) it usually interacts with molecular chaperones which become part of the functional structure if the protein is an enzyme.

Concept questions

(1) Name the four main types of macromolecules in a cell, the units from which they are assembled and how these assembly units are connected.

(2) Discuss the versatility of protein structure; particularly, what enables proteins to exist functionally in an aqueous or hydrophobic environment?

(3) Discuss the processes of enzyme inhibition. Consider kinetic and structural processes.

3

Problems

(1)   You have just received a shipment of guaranteed functional RNAase A, an enzyme that degrades RNA but not DNA and which also does not decompose at boiling temperature. In order to save on your budget, you ordered the cheapest one, which may contain traces of DNAase, a heat-sensitive nuclease. Unfortunately, in your hands the RNAase just would not work, and your DNA preparations contain gobs of RNA. Discuss what your mistake might have been.

(2)   What has to be changed in order to make a cell contain a particular lipid that it does not naturally contain? Discuss two possibilities.

Summary worksheet: fill in the blanks

This introductory chapter aims to remind you that of the four types of cellular macromolecules the .... are the ones that take care of business' in the cell. Consisting of one or more .... chains, the proteins are the direct expression product of the .... information in the nucleic acids. While the nucleic acids function as the keeper of the secret of life, the .... and .... are merely products of specific enzymic functions of proteins. In total, proteins have three distinct functions: (i) they provide .... to the cell as architectural units, (ii) they have .... activity which allows the breakage and formation of bonds between atoms and molecules and (iii) they .... many cellular processes. The latter involves a ....-acting DNA binding protein function which is essential in the direct process of DNA perpetuation (Part 4) and transcription (Parts 3 and 6) and its regulation. Next, certain proteins keep others in a functional .... or help them to mature and reach this functional .... Further, many proteins are involved in the activation and deactivation of other proteins, usually as a result of minor modifications such as (de)phosphorylation or (de)methylation. In that capacity, many proteins contribute to .... which is the initiation of an 'appropriate' cellular response to an external signal. In order to understand or even manipulate these processes, it is necessary to characterize at the molecular level the exact structure of the protein as well as the mode of action.

Additional reading

Chothia, C. (1984). Principles that determine the structure of proteins. *Annu. Rev. Biochem.* **53**, 537–572.

Chapter 2: Compartments

Multiple-choice questions

(1) **Which of the following criteria can be used to distinguish between prokaryotic and eukaryotic cells?**
    (a) whether or not they have a plasma membrane.
    (b) whether or not they have a nucleus (membrane-encased nucleoplasm).
    (c) whether or not they are larger than 1 μm.
    (d) comparison of the sequences of homologous proteins.
    (e) comparison of the signature sequences of 16S rRNA.

(2) **Which of the following structures can be found in both prokaryotic and eukaryotic cells?**
    (a) nucleus
    (b) plasma membrane
    (c) ribosomes
    (d) mitochondria
    (e) cell wall

(3) **There are a few exceptions to the general statement: 'cellular life is impossible without a plasma membrane'.**
    (a) true   (b) false

(4) **Proteins that are embedded in or reach across a membrane are characterized by:**
    (a) a stretch of at least 16 to 21 hydrophobic amino acids in their primary structure.
    (b) a high lysine, arginine and histidine content.
    (c) a high glutamate and aspartate content.
    (d) at least one α-helical or β-sheet domain.
    (e) the C-terminal sequence 'KDEL'.

(5) **The endoplasmic reticulum:**
    (a) extends from the plasma (=cytoplasmic) membrane into the cytoplasm, thereby creating natural entrances utilized in endocytosis.
    (b) extends from the inner nuclear membrane into the nucleoplasm, thereby creating a space that is called the 'nucleolus'.
    (c) extends from the outer nuclear membrane into the cytoplasm, thereby creating a space (lumen) that is separated from the cytosol.
    (d) extends from the outer nuclear membrane into the cytoplasm where it forms the Golgi apparatus.
    (e) is a collective term for the microsomal fraction and consists of lysosomes, peroxisomes and transport vesicles.

(6) **Microtubules extend from so-called microtubule organizing centers (MTOCs). Growth and shrinkage of microtubules are regulated by the free concentration of tubulin and microtubule associated proteins (MAPs).**

(a) true   (b) false

(7) **Eukaryotic cell organelles are circumscribed by two membrane bilayers (envelopes).**

(a) These two membranes have identical composition and are thus indistinguishable.

(b) Because bacterial plasma membranes also consist of two membrane bilayers, it is believed that bacteria and organelles share a common ancestor.

(c) The two membranes differ in their composition which creates an asymmetrical envelope.

(d) The inner and outer membranes have separate functions in emancipation from and communication with the cytosol.

(e) Transport pumps (ATPases) and channels are found solely in the outer membrane.

(8) **The cell division cycle of eukaryotic cells requires that the chromosomes can segregate outside the nucleus. This is accomplished:**

(a) by pre-division transport (prophase) of the chromosomes through the nuclear pores into the cytoplasm.

(b) by an organized breakdown of the nuclear envelope as well as its 'inner lining', the lamina.

(c) by expansion of the nucleoplasm (by uptake of cytoplasm) in that the nuclear membrane utilizes the ER to extend until it 'lines' the plasma membrane; the cell becomes a gigantic nucleus during mitosis.

(d) by the opening of the nucleus on the MTOC side and release of the chromosomes.

(e) very rarely, therefore, cell synthesis occurs predominantly outside of the 'mother' cell and the materials, including chromosomes, have to be transported outside.

(9) **Centrosomes are:**

(a) the entity on the chromosome that guides its movement at mitosis.

(b) the two copies of the chromosome after replication in interphase.

(c) the microtubule-organizing centers during mitosis.

(d) the condensed 'body' of DNA in the nucleus rich in rDNA.

(e) organelles loaded with lytic enzymes.

(10) **The immediate product(s) of the central process in sexual reproduction, meiosis, is (are):**
- (a) the separation of the sex chromosomes during the cell divison cycle.
- (b) the generation of four haploid sex cells or gametes.
- (c) the formation of the zygote, a diploid cell, with a set each of the maternal and paternal chromosomes.
- (d) the arrival of homologous chromosome pairs at the poles, also known as anaphase.
- (e) the generation of a tetraploid cell, in which the $2n$ (diploid) chromosome set has been duplicated.

Concept questions

(1) How is it possible to predict the location and functional structure of a protein from its primary structure?

(2) Name the essential cellular structures and how they contribute to cellular function.

(3) How does the cell create asymmetric membranes, the prerequisite for a functioning bioenergetic machinery in every cell?

(4) Discuss the transport of a protein from its producer, the ribosome, to its final destination as an extracellular matrix protein.

Problems

(1) An economically valuable protein would become more efficient if it could be strictly localized at the inner surface of the plasma membrane. Devise a strategy that will accomplish just that.

(2) In an adult organism, the cells of tissues and organs that are not continuously regenerating are terminally differentiated. Discuss the characteristics that distinguish these healthy cells from tumcrous cells and propose a generalized therapy to stop tumor proliferation.

Summary worksheet: fill in the blanks

There are various ways of classifying all living organisms and structural, physiological, biochemical and molecular genetical criteria have all been used. They all support a fundamental distinction between prokaryotic and eukaryotic cells; however, there are common features to both which are what we know as the 'hallmarks of (cellular) life'. Both types of cells comprise a distinct compartment that is always separated from the chaotic

7

environment by membranous .... While prokaryotic cells can be considered a single compartment, a eukaryotic cell consists of many subcompartments including douple membrane-enveloped organelles, the endoplasmic reticulum and Golgi apparatus. Furthermore, inside and outside of the eukaryotic cell are structures collectively called the .... that are involved in the organization of the cell's structure and function. From the standpoint of molecular genetics, these are not merely additional compartments and structures; instead, they contribute intimately to the complex processes of the maintenance and replication of .... information as well as gene .... and protein .... in the cell. In that regard, it is crucial to understand cellular function at the macroscopic level (e.g., membrane fusion, microtubule assembly, nuclear breakdown or chromatin condensation). In addition, an understanding of both the prokaryotic and eukaryotic cell division cycles is crucial for the understanding of individual events and processes in the field of molecular genetics. While the karyotype is maintained in somatic cells through ...., ... eukaryotic cells can reproduce also sexually. This requires gametogenesis which is implemented through .... Meiosis and the fertilization process carry out the Mendelian segregation of traits.

Additional reading

Davis, L.I. (1995). The nuclear pore complex. *Annu. Rev. Biochem.* **64**, 865–896.

McIntosh, J.R. and McDonald, K.L. (1989). The mitotic spindle. *Scientific American*, October.

Robinow, C. and Kellenberger, E. (1994). The bacterial nucleoid revisited. *Microbiol. Reviews* **58**, 211–232.

Saier, M.H. Jr (1997). Peter Mitchell and his chemiosmotic theories. *ASM News* **63**, 13–21.

Part 1: DNA as information

Theme

Part 1 covers the structure of nucleic acids in more general terms from two points of view: what does the structure represent in terms of inheritable information and how can the structure be investigated? You will be reminded that the first decades of this century produced significant knowledge about the structural complexity and diversity of structural proteins and enzymes. This, unfortunately, turned out to be the major obstacle to identifying and accepting nucleic acids as the molecular carrier of genetic information. DNA structure appeared too simple to 'contain' the necessary wealth of information. An unconsidered fact was that the carrier of genetic information needed to adhere to an unfailing principle that guaranteed the faithful maintenance of this information. We know now that this is exactly embodied in the semiconservative replication (a template–replica relationship) of usually double-stranded DNA which contains the inheritable information in form of a defined base sequence. This nucleotide sequence is translated into amino acid sequence according to a universal code, the genetic code. A selective utilization of the genetic information requires that the DNA duplex structure is flexible in terms of separation (melting) and renaturation (annealing). This is accomplished by the unique chemical properties of DNA as well as a diverse toolkit of DNA-modifying enzymes. The *in vitro* utilization of many DNA-modifying enzymes such as endonucleases and polymerases and our present knowledge about general physical and chemical properties of DNA has recently enabled researchers to unravel positional and physical information for entire genomes. Techniques like DNA fingerprinting have enhanced the quality of public life tremendously through their contribution to diagnosis of disease and to forensic science. Your knowledge of the principles introduced in Part 1 will make it easy to understand the functional aspects of nucleic acids as they are discussed in Parts 2 to 7.

General reading

Hartl, D.L. (1994) *Genetics*, 3rd edition. Jones & Bartlett, Boston, London.
Stryer, L. (1995) *Biochemistry*, 4th edition. W. H. Freeman & Co., New York.

Key terms

Use the following key terms as a guide for your study and the composition of the summary at the end of each chapter.

Chapter 3

- allele
- chromosome
- cistron
- complementation group
- complementation in *trans*
- gain-of-function mutation
- gene
- genetic loci
- genome
- genotype
- inheritance
- linkage group
- phenotype
- polymorphism
- reciprocal recombination
- transcriptional unit
- wild-type

Chapter 4

- adenine
- amino acids
- antiparallel
- codon
- complementary
- cytosine
- deletion
- deoxyribonucleic acid
- genetic code
- glycosidic bond
- guanine
- hydrogen bond
- insertion
- nitrogenous bases
- nucleic acids
- nucleotide
- nucleotide sequence
- phosphodiester bond
- point mutation
- polynucleotide
- protein
- semiconservative
- thymine
- trinucleotide

Chapter 5

- annealing
- complementary
- DNA melting
- double helix
- hybridization
- major groove
- palindrome
- single-stranded
- stem–loop
- supercoiling
- temperature-dependence
- topoisomerases

Chapter 6

- amino acid
- colinearity
- DNA sequencing
- endonuclease
- eukaryote
- exon
- gel electrophoresis
- intron
- linkage
- nucleoside
- nucleotide sequence
- phenotype
- prokaryote
- physical map
- restriction fragment length polymorphism

Chapter 3: Genes are mutable units

Multiple-choice questions

(1) **What is the 'genome'?**
   (a) the dwarf phenotype of an organism.
   (b) the total content of gene-bearing molecules in an organism.
   (c) the number of chromosomes in a diploid cell.
   (d) the unit of inheritance.
   (e) the total content of gene-bearing molecules in a cell specific to the organism.

(2) **The phenotype of a homozygote directly reflects the genotype of an individual allele. Therefore merodiploid bacteria rarely differ in phenotype from that of the wild-type.**
   (a) true   (b) false

(3) **The F_2 progeny of an animal shows a distinct distribution of parental traits, with 25% displaying the original paternal phenotype, 25% the original maternal phenotype and 50% displaying both parental traits. A Mendelian analysis allows classification of such a pattern of inheritance (segregation of alleles) as:**
   (a) complete dominance
   (b) partial dominance
   (c) codominance
   (d) linkage
   (e) reciprocal recombination

(4) **The colony morphology (phenotype) of a mutant bacterium is indistinguishable from wild type. The mutation is probably:**
   (a) a point mutation
   (b) a nonsense or missense mutation
   (c) a third codon base substitution
   (d) (a) and (b)
   (e) (a) and (c)

(5) **Identify all incorrect statements:**
   (a) The smaller the proportion of recombinants in the progeny the tighter the linkage of traits (genes).
   (b) Differences between parental and progeny chromosome structures ('shape') indicate the formation of genetic recombinants (physical crossing-over) during meiosis.
   (c) If the number of F_2 recombinants is maximal for two given original parental traits they are considered tightly linked and the respective loci are very close to one another on the same chromosome.

11

(d) The maximum recombination between two loci on a single chromosome is 50% because only two of the four associated chromatids participate in an individual recombination event.

(e) Mutational analysis can be employed to show that the gene itself does not necessarily have a linear construction (introns !) while the array of genes on a chromosome is linear.

(6) Consider a missense mutation occurring in a diploid yeast cell. The mutation causes the introduction of a different amino acid into the catalytic site of a catabolic enzyme which allows the enzyme to utilize additional substrates. This so called 'gain of function mutation' is always dominant.

(a) true   (b) false

(7) When the 'gene' was finally established as the unit of inheritance by mutational analysis, a remaining task was the assignment of a functional value at the molecular level. This was satisfactorily resolved with the finding that:

(a) A mutation in a dominant allele can cause it to behave like a recessive allele.

(b) A mutant gene fails to produce a protein found in the wild-type organism.

(c) A mutant gene fails to produce a polypeptide chain identical to the one found in the wild-type organism.

(d) Auxotrophic bacteria can grow in unsupplemented medium only after exposure to a mutagenic compound.

(e) The heterozygote resulting from mating of a mutant phenotype and a wild-type gamete always displayed the wild-type phenotype.

(8) In your attempt to characterize a particular gene you have finally succeeded in producing a selectable bacterial mutant. However, to claim fame for the identification of the gene you have to prove that the isolated gene solely contributed to the observed mutant phenotype by:

(a) exposing the bacteria to mutagenic compounds and isolating a revertant to the wild-type phenotype.

(b) failure to obtain complementation in *trans* in that transformation of the mutant bacterium with a plasmid that carries the mutant gene will not restore the wild-type.

(c) failure to obtain complementation in *trans* in that transformation of the wild-type bacterium with a plasmid that carries the mutated gene will not produce the mutant type.

(d) complementation in *trans* in that transformation of the mutant bacterium with a plasmid that carries the mutated gene produces the wild-type phenotype.

(e) complementation in *trans* in that transformation of the mutant bacterium with a plasmid that carries the wild-type gene restores the wild-type phenotype.

12

(9) **Polymorphism (detectable by phenotypic criteria or DNA analysis) is:**

    (a) the occurrence of different alleles in a culture of a single colony-purified bacterium.

    (b) the occurrence of at least two different alleles in a population of one species.

    (c) the occurrence of at least three different alleles in a population of one species.

    (d) the situation in which a gene affects two or more unrelated aspects of a phenotype.

    (e) the situation in which a cell contains more then two sets of the haploid genome.

(10) **When two mutated segments of DNA fail to complement one another in *trans* it can be inferred that both mutations affect the same function. Both of these mutations and each further mutation that fails to complement in *trans* are assigned to the same complementation group and are believed to be part of a discrete genetic unit. This unit may be a cistron or it may be a multicistronic transcriptional unit if the mutation reliably interferes with the process of transcription.**

    (a) true   (b) false

Concept questions

(1) What is the phenotype of an organism?

(2) What is a linkage group? Give an example of linked loci.

(3) What is a 'cistron'? Discuss the terms 'gene' using 'complementation' and 'allele'?

(4) Give an example of a 'gain-of-function' mutation.

(5) Discuss 'polymorphism' by using an example (e.g. Landsteiner bloodgroups).

Problems

(1) Why is transformation/conjugation/transfection necessary for performing a complementation (in *trans*) analysis in bacteria? Also discuss/correlate the terms: chromosome, genome, gene, trait, cistron, transcriptional unit, translational unit and ploidy.

(2) You have identified a case in which peculiar events during storage of seeds used for breeding have rendered rice plants highly susceptible to a bacterial pathogen. Your hypothesis is that a chemically-induced mutation has caused the loss of a crucial enzyme with antibacterial function. Design an experiment (by utilizing your knowledge about complementation analysis) with which you can test your hypothesis and

simultaneously work out a strategy to regain resistance. Hint: the bacterium *Agrobacterium tumefaciens* has the unique capability of incorporating bacterial DNA into the plant genome by means of its Ti plasmid.

Summary worksheet: fill in the blanks

It has been known since the experiments of .... that .... segregate to the progeny independently in a predictable way (following Mendel's laws). These traits are the visible results of the genetic activity of Mendel's 'discrete factors', the ...., which are the structural units of .... Thus genes assort independently like the traits. Diploid cells contain at least two variants of a gene called .... which are part of different .... homologs, hence .... segregate following the pattern of chromosome distribution (in mitosis and meiosis). Genetic .... (or specific alleles) on a chromosome belong to a specific .... group and are usually inherited together if they are physically close to one another. Chromosomes are not equal to .... groups because of the crossing-over (= ....) during meiosis in which chromosome translocation occurs.

Each cell or multicellular organism has a defined and characteristic .... which provides the blueprint for its morphological appearance, the .... If this is the phenotype/genotype that has been first recognized for a particular species it is called the .... Any deviation from the .... is considered a mutant phenotype/genotype. A mutation is the change in the sequence of DNA. If a population of a species contains more than ...... variants of a gene there is the problem that some species may be considered mutant phenotypes when compared with the wild-type phenotype. Because determination of the original variants of the species is difficult such a population 'genotype' is called .... This provides an evolutionary context to the change of genetic information by mutation in which .... is resolved by fixation (creating a gene family) or loss. Both contribute to speciation and the presence of nonfunctional or pseudogenes in many recent genotypes.

While the gene was defined as the unit of ...., the .... is the equivalent term for the unit of genetic function. The .... codes for a single polypeptide chain that may be sufficient to constitute an

14

enzyme or structural protein. It can be identified by .... of a mutation considering the ploidy (diploid cells usually have a mutant/wild-type genotype after mutagenesis) and genetic nature (eukaryotic or prokaryotic) of the cell. If mutations fail to complement in .... they are assigned to the same .... You will learn in Parts 2 and 3 that prokaryotic transcripts are usually polycistronic. Consequently a(n insertional) mutation in the first cistron may affect transcription of the entire transcriptional unit and the complementation group will extend from a cistron to the entire .... In contrast, a eukaryotic cistron corresponds to a complementation group.

Additional reading

Cairns, J. (1990). The origin of mutants. *Nature* (London) **335**, 192–194.

Chapter 4: DNA is the genetic material

Multiple-choice questions

(1) **Two experiments were crucial for the identification of DNA as the genetic material: a study of bacterial (*Pneumococcus*) virulence in mice and the infection of *Escherichia coli* by phage T2. The major argument in both cases was that:**
   (a) DNA was reisolated from the infected organisms as the 'causal agent' of disease.
   (b) Mutated DNA caused a failure of virulence.
   (c) DNA—but not protein—taken up from an external source by an organism changed its genetic potential.
   (d) DNA was not transferable between organisms and hence must be a very conservative molecule.
   (e) DNA from prokaryotes, eukaryotes and viruses could be mixed and substituted for one another.

(2) **ATP used hydrolytically for the catalysis of anabolic reactions and the ATP incorporated into a nucleic acid are different molecules.**
   (a) true   (b) false

(3) **A nucleic acid:**
   (a) consists of a nitrogenous base, a pentose sugar and a phosphate group.
   (b) is a polymer of nucleosides linked at the 3' and 5' positions of the pentose rings.
   (c) is a polymer of nucleotides linked by 5'–3' phosphodiester bonds.
   (d) is a double strand of polynucleotide chains.
   (e) is a double helix of antiparallel polynucleotide strands.

(4) **Watson and Crick proposed in 1953 that:**
   (a) polynucleotide DNA chains associate in a double helix by hydrogen bonding.
   (b) replication of DNA is semiconservative, always creating parent– daughter duplexes.
   (c) the genetic code is represented by successive trinucleotides.
   (d) DNA and not RNA is the universal genetic material.
   (e) isolation of revertants can prove that a mutation is not a deletion mutation.

(5) Replication of DNA:
- (a) involves the synthesis of two daughter strands which associate in a double helix.
- (b) follows the principle that a new daughter strand is assembled on each parental strand.
- (c) is dependent on the species-specific genetic code.
- (d) is the most common source for base mispairing.
- (e) is a process that describes gene expression.

(6) A mutation outside the coding region does not cause a change in the phenotype of the cell or organism.
- (a) true   (b) false

(7) Which of the following statements is incorrect?
- (a) 20 different codons represent the genetic code.
- (b) Tryptophan and methionine are encoded by just one codon.
- (c) Every nucleotide triplet encodes an amino acid.
- (d) Different codons may code for the same amino acid.
- (e) The third position in a codon is highly variable.

(8) Which base pairs are found in double-stranded DNA?
- (a) A•U   (b) G•T   (c) C•G
- (d) T•A   (e) C•A

(9) Spontaneous mutations are expected to occur randomly but equally distributed across the genome, however, detailed analysis shows that some locations in the DNA are more susceptible to mutation than others. This so-called 'hotspot' mutation is due to:
- (a) the spatial structure of DNA which exposes only selected portions of the DNA to mutagenic effector molecules.
- (b) the existence of modified (e.g. methylated) bases which favor further modification such as deamination that lead to base transition and after replication to a muation.
- (c) the existence of modified (e.g. methylated) bases which favor mismatch replication and hence the creation of mutations.
- (d) the existence of DNA segments rich in A-T which allows for spontaneous melting and the incorporation of base-mismatching nucleotides.
- (e) the low pressure against silent mutations, hence hotspots are always constituted by the third base in a codon.

(10) Mutations that occur in other genes and restore the wild-type from a mutant phenotype (= circumvention of a mutation in the original gene) are called repressor mutations.
- (a) true   (b) false

17

Concept questions

(1) Why is DNA exceptionally suited to be the genetic material?

(2) Name an example where DNA does not function as the 'transforming principle of inheritance'.

(3) What is the difference between base pairs with regard to their biochemistry and information?

(4) Under what conditions is it possible to predict the percentage of 'G' in a given nucleic acid?

(5) Explain the following terms: mutant; mutation; mutagen; mutagenesis. Characterize point, insertion, deletion, and suppression mutations in context with their effects on phenotype.

Summary worksheet: fill in the blanks

One of the most important and exciting discoveries in the history of science was that inheritance can be attributed to a single kind of molecule, the .... The time between 1928 and 1952 was filled with many ingenious experiments that tried to prove that the transforming principle is .... The realization that the content of eukaryotic chromosomes, the transfecting principle in eukaryotic cells, the transforming principle in bacteria and the transduction of phage material can be reduced to the same molecule finally established that nucleic acid is the universal genetic material. Theoretically the major obstacle to accepting this fact was the very .... of nucleic acids which was known to be a .... chain. DNA, the genetic material in all prokaryotic and eukaryotic cells, exists as an .... double strand of 2-deoxyribose molecules linked by 5'-3' .... bonds. The genetic information resides solely in the kind and sequence of .... that are linked to position 1 of the ribose by a .... bond. The structure of the bases dictates that each of them pairs with only one other naturally: the purine .... (G) pairs with the pyrimidine .... (C) and the purine .... (A) pairs with the pyrimidine .... (T). Because a G•C pair establishes three .... bonds G•C pairs are more stable than A•T pairs which involve only two hydrogen bonds. Bases that can pair are called .... bases. This structure seemed to be too simple to contain all the information for the complicated structure of a cell. This obvious

18

discrepancy was resolved by understanding two further mysteries: (i) how can DNA code for the incredible diversity of .... structures and (ii) how can the genetic information be maintained faithfully over generations? The solution to the first mystery was the finding that the genetic information stored in the bases and their sequence (= the .... sequence) of selected segments of the DNA is 'translated' into a sequence of .... in a polypeptide following a universal code, the .... This code reliably assigns .... to a specific amino acid. The polypetides in turn assemble to and/or function as structural proteins or enzymes both of which contribute to the maintenance and complexity of a cell. The second mystery was resolved by the discovery that the replication of DNA is .... or that 'new' is always paired with 'old'. This means that daughter cells always start their independent cellular life with a complete genome consisting of a .... strand and a new complementary strand of DNA. By taking this into account it is a next logical step to explain variation between cells and organisms with distinct changes in the .... These changes can be generated by the substitution of a single base called .... or by the .... or .... of one to thousands of nucleotides.

Additional reading

Farabaugh, P.J. (1996). Programmed translational frameshifting. *Annu. Rev. Genet.* **30**, 507–528.

Ripley, L.S. (1990). Frameshift mutation: determinants of specificity. *Annu. Rev. Genet.* **24**, 189–213.

Chapter 5: Nucleic acid structure

Multiple-choice questions

(1) **Melting or denaturation of the DNA duplex disrupts hydrogen bonds between complementary bases and hence changes their absorption characteristics. The DNA melting temperature:**

    (a) is approximately 45°C for mammalian DNA therefore fever above 42°C is extremely dangerous.

    (b) depends on the A•T content because the more A•T is present the less energy is needed to separate the two strands.

    (c) can be defined as the midpoint of the temperature range over which the DNA strands separate.

    (d) can be determined by the change in absorption at 260 nm which is characteristic for the bases.

    (e) is the temperature at which single-strand breakage occurs (phosphodiester bonds break).

(2) **Hybridization of DNA single strands to form a duplex under high salt and low temperature conditions indicates nearly perfect complementarity and the process can be considered a renaturation (annealing) reaction.**

    (a) true   (b) false

(3) **Denaturation of DNA:**

    (a) can involve the melting of the duplex

    (b) can be caused by lowering of the temperature

    (c) is reversible

    (d) is the breakage of phosphodiester bonds

    (e) can involve the disruption of hydrogen bonds

(4) **Hairpin formation in nucleic acid duplexes such as DNA does not occur as frequently as in single-stranded molecules. It requires the existence of a palindromic sequence which allows the two strands to form symmetrical hairpins that appear as a cruciform structure.**

    (a) true   (b) false

(5) **Single-stranded nucleic acids such as RNA display so-called secondary structures. This hairpin formation:**

    (a) is based on the complementarity of individual segments allowing the formation of antiparallel duplexes.

    (b) depends on the content of A•U because the fewer hydrogen bonds have to form the less energy is necessary for the base pairing to occur.

    (c) occurs only if all bases in the two pairing segments are complementary to one another.

  (d) involves also irregular base pairings such as G•U.

  (e) allows the participation of several bases that cannot base-pair as long as the overall free energy necessary for the base-pairing is positive.

(6) While the B-form of the double helix represents the general structure of DNA, the Z-form has been identified so far only in some lower eukaryotic cells.

  (a) true   (b) false

(7) Supercoils in a DNA molecule:

  (a) occur only in circular DNA. If the duplex is closed after it has been twisted around its own axis (=increasing the writhe) the duplex becomes arrested under torsional pressure.

  (b) occur in circular and linear DNA. The increase in writhe can be arrested by changes in the base-pairing and additional hydrogen bonds.

  (c) can constitute an underwound duplex in a closed DNA. Thus negative supercoiling is a precondition for DNA modifying activities to provide access to the enzymes.

  (d) are the reason for the condensation of eukaryotic DNA during mitosis.

  (e) are the sum of the rotation of the strands around one another within the duplex and the turning of the axis of the duplex in space.

Concept questions

(1) Name the conditions that facilitate DNA renaturation (annealing).

(2) Give an example of the importance of stem–loop structures formed in single-stranded nucleic acid.

(3) Why is it important that the DNA douple helix maintains defined grooves?

(4) Depict the structure of a nucleoside and a nucleotide and discuss the difference between DNA and RNA with regard to their stability.

Problems

(1) You have obtained a sample of a negatively supercoiled circular DNA and you want to perform an *in vitro* transcription experiment. Your first experiment carried out on the bench does not produce satisfactory results. Discuss possible ways to improve your experiment.

(2) The polymerase chain reaction involves melting and annealing between a primer and a DNA template strand. Discuss the impact of the temperature cycling regime on the stability of the primer–DNA duplex and thus on PCR yield.

Summary worksheet: fill in the blanks

While the structures of polynucleotide chain duplexes are already fairly rigid, their organization in a .... and the additional twisting of the duplex around its axis (........................) may seem to 'freeze' the nucleic acid into a completely inflexible molecule. This, however, is not the case for several reasons: first, the structure of a double-stranded nucleic acid is strictly dependent on several environmental conditions including ...., pH value and the concentration of salts (e.g. magnesium ions) and it will thus change with fluctuations in these parameters. Second, the cell has a toolkit of enzymes which modify structure and topology of DNA such as .... Third, but not last, the DNA of a circular or linear chromosome/genome can fit into tight spaces and endure mechanical stresses. The complex structure of the DNA duplex molecule represents a successful adaptation that allows for faithful replication and the organized utilization of the genetic information. The ability of .... segments within a .... nucleic acid to form partial duplex structures accounts for .... structures. These hairpin or .... structures play a role in the regulation of DNA modification processes in that they influence the performance of polymerases and other DNA-binding proteins. In Parts 2 to 7 you will learn much about how particular nucleic acid structures influence the maintenance and retrieval of genetic information.

Additional reading

Dickerson, R.E., Drew, H.R., Conner, B.N., Wing, R.M., Fratini, A.V., and Kopka, M.L. (1982). The anatomy of A-, B-, and Z-DNA. *Science* **216**, 475–485.

Frank-Kamenetskii, M.D. and Mirkin, S.M. (1995). Triplex DNA structures. *Annu. Rev. Biochem.* **64**, 65–95.

Gold, L., Polisky, B., Uhlenbeck, O., and Yarus, M. (1995). Diversity of oligonucleotide functions. *Annu. Rev. Biochem.* **64**, 763–797.

Rennie, J. (1993). DNA's new twists. *Scientific American*, March.

Schleif, R. (1992). DNA looping. *Annu. Rev. Biochem.* **61**, 199–223.

Chapter 6: Isolating the gene

Multiple-choice questions

(1) **Analysis of the length of DNA fragments created by digestion of double-stranded DNA with selected restriction endonucleases is suited to physical mapping because:**
   (a) The linearity of DNA allows immediate determination of the restriction fragment sequence even if just one restriction enzyme is used.
   (b) Endonucleases cut DNA at equidistant sites giving fragments of a length specific to the organism.
   (c) The linearity of DNA means that the lengths of the restriction fragments add up to the total length of the DNA whereby overlapping fragments resulting from double cuts create an unequivocal map.
   (d) The digestive activity of these endonucleases is restricted in a species-specific manner.
   (e) This technique simultaneously provides extensive information on the nucleotide sequence.

(2) **Restriction fragment analysis allows exact analytical comparison of alleles because:**
   (a) endonucleases cut only one of the two variants.
   (b) it utilizes restriction sites as genetic markers.
   (c) alleles in a given cell are never identical regarding their nucleotide sequences.
   (d) it is independent of the function of the products of the variants.
   (e) the cutting sites of endonucleases are restricted to coding regions in the DNA.

(3) **Restriction analysis of end-labeled DNA allows exact identification of the restriction site nearest to the unlabeled end.**
   (a) true   (b) false

(4) **Restriction fragment length polymorphism (RFLP) is:**
   (a) the technique used to 'fingerprint' patterns of inheritance.
   (b) the difference in the restriction maps between the two alleles in a diploid cell.
   (c) the difference in the restriction maps between two individuals of one species.
   (d) the difference in the restriction maps between two individuals of two species.
   (e) the difference in the restriction maps of haplotypes for two different enzymes.

(5) **A restriction map and an RFLP map significantly different in the first one is a physical map while the latter is a linkage map.**

(a) true   (b) false

(6) **In order to construct an exact physical map of a eukaryotic genome:**

(a) it must be possible to separate the individual chromosomes cleanly (which is achieved by pulse field gel electrophoresis).

(b) DNA must be isolated from haploid cells (gametes) only. This avoids ambiguous information due to the existence of homologous chromosomes in diploid cells.

(c) single chromosome analysis must be performed, so analysis is possible during anaphase only.

(d) it is necessary to find restriction enzymes that cut only once per chromosome.

(e) it is necessary first to separate nuclear and organellar DNA and then to subject them to individual mapping.

(7) **There are two different ways to determine the nucleotide sequence of a nucleic acid: the chemical sequencing (Maxam–Gilbert) method and the enzymatic sequencing (Sanger) method. The basic principle/advantage of the Sanger method is:**

(a) the differential interaction of the bases with particular dyes.

(b) extension of a synthetic primer and reliable termination of DNA repair synthesis.

(c) the correlation of restriction sites with the end-label of the DNA.

(d) the ability to 'sequence' both strands of the DNA duplex simultaneously.

(e) that only DNA but not RNA will be identified. This allows investigation of less pure preparations and so reduces costs.

(8) **Prokaryotic genes and proteins are said to be 'colinear'. This:**

(a) refers to the fact that the primary transcript is also the translatable mRNA.

(b) refers to the fact that the ribosome can translate a coding DNA region directly.

(c) allows 'reverse genetics', meaning that it is possible to construct primers and probes for the identification of particular genes based on the amino acid sequence of the respective protein.

(d) refers to the fact that the length of the mRNA defines the length of the synthesized protein.

(e) means that the number of deoxyribonucleotides in a coding region is exactly three times larger than the number of amino acids in the respective polypeptide chain.

24

(9) **Eukaryotic genes are often interrupted. This:**
   (a) reflects the fact that eukaryotic mRNAs are polycistronic.
   (b) refers to the separation of coding units, the 'exons', by noncoding 'introns'.
   (c) refers to the fact that eukaryotic DNA is linear and separated in individual chromosomes. Hence genes may be partly on one chromosome and partly on another.
   (d) means that the primary transcript must be processed in order to be translatable.
   (e) means that eukaryotic genes may have multiple expression products because it is possible to use different combinations of exons to construct the mRNA.

(10) **The genetic element of a virus can contain between one and 300 individual genes. In contrast to living organisms, the genetic element of a virus can be DNA or RNA (never both!), hence DNA is not the truly universal genetic material.**
   (a) true   (b) false

Concept questions

(1) Describe the use of endonucleases for physical mapping of unknown DNA. What basic feature of the gene makes restriction analysis suitable for mapping?

(2) What is restriction fragment length polymorphism (RFLP)?

(3) Explain the methodological principle employed for Sanger sequencing of DNA.

(4) Discuss three ways in which a given DNA sequence codes for more than one protein.

(5) Discuss why the 'life' cycle of RNA viruses such as the influenza virus constitute an excellent support for the statement that viruses are just 'parasitic (infectious) genetic elements' rather than organisms.

Problems

(1) A mutation abolishes the synthesis of a monomeric protein in a diploid yeast cell. Discuss the location of the mutation considering that both alleles in the cell seem to be nonfunctional.

(2) It has just been disclosed in the journal *Science* that a major inheritable disease can be artificially induced *in vitro* by mutagenic agents that cause base substitutions in DNA. Develop a quick diagnostic strategy that allows identification of the disease genotype.

25

Summary worksheet: fill in the blanks

The emphasis of Part 1 was to introduce the basic structure and function of nucleic acids, the universal carrier of inheritable information. Chapter 6 in particular introduced principles and methods that allow characterization of nucleic acids in terms of a combination of .... dissection (nucleic acid sequence) and .... information (localization of individual genes). Nucleic acid analysis is based upon two major principles: the specificity of .... for particular recognition sites and the template- and primer-dependence of nucleic acid .... While this apears trivial in retrospect, the application of these principles to analytical investigation of DNA represented a novel integration of many not universally accepted hypotheses. Both basic techniques produce nucleic acid fragments of different .... which are usually separated by agarose .... The negative charge of DNA is proportional to its .... due to fact that the .... are polymerized via phosphodiester bonds hence electrophoretic separation occurs strictly by fragment size. This allows .... The length of a DNA fragment can be determined by using known DNA fragments such as a restriction digest of lambda phage DNA selected for by the desired range of fragment length resolution and running sample and standard side by side on a gel. The fact that .... action is sensitive to mutation in the restriction site allows variant analysis independent of the .... A mutation that abolishes or creates a restriction site constitutes a genetic polymorphism that is called .... (RFLP). RFLP analysis of chromosomes creates a .... map while determination of the .... establishes a precise physical map. Both techniques have contributed to our understanding that the primary structures of DNA (.................... sequence) and protein (.................... sequence) are colinear in .... but not ....

Additional reading

Bauer, W.R., Crick, F.H.C., and White, J.H. (1980). Supercoiled DNA. *Scientific American*, July.

Hiraga, S. (1992). Chromososme and plasmid partitioning in *Escherichia coli. Annu. Rev. Biochem.* **61**, 283–306.

Wake, R.G. and Errington, J. (1995). Chromosome partitioning in bacteria. *Annu. Rev. Genetics* **29**, 41–67.

White, R. and Lalouel, J.-M. (1988). Chromosome mapping with DNA markers. *Scientific American*, February.

Part 2: From gene to protein

Text chapters

Theme

The seven chapters of Parts 2 and 3 and the six chapters of Part 6 describe how the genetic information stored within the primary structure of DNA is exploited and used for the sustenance of life, a complex process usually referred to as 'gene expression'. While Part 3 (for prokaryotes) and Part 6 (for eukaryotes) are concerned with the processes, mechanisms and their regulation of the selection of the genetic information necessary to maintain cellular vitality, Part 2 covers the second stage of gene expression: the mechanisms of translation of the primary transcript into polypeptides. This translation is based upon the genetic code and catalyzed by the ribosome, a complex consisting of rRNA associated with various polypeptides. The polymerization of amino acids, however, does not necessarily yield a functional product. Hence, polypeptide synthesis is usually followed by extensive processing, targeting and folding ensuring that the polypeptide ends up at the location of its function as a structural or catalytically active protein.

A justification for dealing with the second step of gene expression first may be found in the fact that the products of translation, polymers of amino acids held together by peptide bonds, are structurally similar in prokaryotic and eukaryotic cells. Nevertheless, there are significant differences between translation in prokaryotic and eukaryotic cells: First, the substrate for protein synthesis (the template for translation), the messenger RNA (mRNA), has a different structure. Therefore, second, the initiation of translation differs between prokaryotes and eukaryotes. Third, the translation complex of prokaryotes differs from that in the cytoplasm of eukaryotic cells, but is similar to the eukaryotic organellar translation machinery. Also, because of extensive compartmentalization, the process of translation in eukaryotes is completely uncoupled from the processes of DNA replication (Part 4) and transcription (Part 3). Organelles such as mitochondria and chloroplasts have maintained organelle-specific genetic material, however, its expression is under metabolic control of the nucleus. In addition, eukaryotic cells synthesize polypeptides at two different locations in the cytoplasm: the products of rough endoplasmic reticulum (RER) and cytoskeleton-bound ('free') ribosomes are destined for 'external' (cell membrane integration, secretion) and 'internal' affairs (cytoplasmic proteins, transport into cell organelles), respectively. More complicatedly, the cells of multicellular eukaryotic organisms, whether organized as single cells, tissues or organs, can assume and utilize their specific functional fates only through the differential utilization of genetic information in the individual cells.

The crucial macromolecule in the process of translation is ribonucleic acid (RNA) which participates in three different forms as ribosomal RNA (rRNA), transfer RNA (tRNA) and mRNA. All three RNA molecules perform different but cooperative functions in preparation for and during translation which also involves many ancillary reactions catalyzed by enzymes. The polymerization reaction (transpeptidation), however, is currently unknown. Intriguingly, some experiments suggest that the process could be facilitated by catalytic rRNA. This unanticipated catalytic property of RNA is generally and extensively discussed in Chapters 30 and 31 of Part 6.

General reading

Neidhardt,F.C. *et al.* (1989). *Escherichia coli and Salmonella typhimurium—cellular and molecular biology*. Volume 2, American Society for Microbiology, Washington, D.C.

Stryer,L. (1995). *Biochemistry*, 4th edition. W. H. Freeman & Co., New York.

Key terms

Use the following key terms as a guide for your study and the composition of the summary at the end of each chapter.

Chapter 7

❑ acceptor arm	❑ endonuclease	❑ primary structure
❑ aminoacyl-tRNA sythetase	❑ exonuclease	❑ ribonuclear protein
	❑ leader	❑ ribosome
❑ anticodon	❑ mRNA	❑ Shine–Dalgarno sequence
❑ A site, P site, E site	❑ nascent polypeptide	
❑ cap	❑ poly(A) tail	❑ trailer
❑ cistron	❑ polycistronic transcript	❑ tRNA
❑ cloverleaf structure	❑ polyribosome	

Chapter 8

- ❑ A site, P site, E site
- ❑ binary complex
- ❑ deacylation
- ❑ elongation factor, EF
- ❑ EF-Ts, Ef-Tu
- ❑ f-Met-tRNA
- ❑ initiation factor, IF
- ❑ IF-1, IF-2, IF-3
- ❑ mRNA

- ❑ nascent protein
- ❑ open reading frame
- ❑ peptidyl transferase
- ❑ peptidyl-tRNA
- ❑ release factor, RF
- ❑ ribosome
- ❑ rRNA
- ❑ 5S, 5.8S, 18S, 28S rRNA

- ❑ 5S, 16S, 23S rRNA
- ❑ r-proteins
- ❑ RBS
- ❑ Shine–Dalgarno sequence
- ❑ stop/termination codons
- ❑ ternary complex
- ❑ translocation
- ❑ tRNA

Chapter 9

- ❑ cognate tRNAs
- ❑ frameshift suppression
- ❑ missense

- ❑ nonsense
- ❑ nucleotide codon
- ❑ readthrough

- ❑ suppressor
- ❑ third base degeneracy

Chapter 10

- ❑ aminopeptidase
- ❑ anchor
- ❑ chaperone
- ❑ co-translational
- ❑ post-translational

- ❑ Sec proteins
- ❑ signal peptide
- ❑ signal recognition particle (SRP)
- ❑ stop-transfer

- ❑ sorting
- ❑ targeting

Chapter 7: Messenger RNA

Multiple-choice questions

(1) **A prokaryotic messenger RNA contains several characteristic segments that are necessary for its function. These are:**
- (a) promoter, Shine–Dalgarno sequence, start codon, stop codon, stem–loop.
- (b) promoter, transcriptional start point, leader, ORFs preceded by a Shine–Dalgarno sequence and separated by intercistronic sequences, trailer, stem–loop.
- (c) transcriptional start point, ORFs preceded by a Shine–Dalgarno sequence and separated by intercistronic sequences, trailer, stem–loop.
- (d) transcriptional start point, leader, ORFs preceded by a Shine–Dalgarno sequence and separated by intercistronic sequences, trailer.
- (e) transcriptional start point, leader, one Shine–Dalgarno sequence followed by ORFs separated by intercistronic sequences, trailer.

(2) **The 'template' or 'antisense' DNA strand can be characterized as follows:**

The template strand is read by RNA polymerase to synthesize a complementary nucleic acid, the mRNA, which is the 'sense'-making template in ribosomal protein synthesis.
- (a) true   (b) false

(3) **The length of a eukaryotic pre-mRNA is determined by:**
- (a) a DNA stemloop sequence which produces an intrinsic terminator RNA structure in the trailer.
- (b) the polyadenylation site at its 3' end at which the transcript is cleaved and extended with poly(A).
- (c) termination proteins which bind to RNA polymerase II at the termination site.
- (d) exonucleases which 'trim' all primary transcripts to final length pre-mRNAs.
- (e) capping and polyadenylation as well as intron splicing.

(4) **Most prokaryotic mRNAs are readily degraded from 3' to 5' before transcription and translation have been completed, hence, mRNA half life is approximately 15 minutes.**
- (a) true   (b) false

(5) **Identify the correct statements about mRNA degradation.**
- (a) Endonucleolytic cleavage of prokaryotic mRNA proceeds 5'–3' behind the ribosomes.

(b) Prokaryotic mRNA degradation occurs always 3'–5' by exonucleases.

(c) Degradation of eukaryotic mRNAs is dependent on the presence of multiple specific destabilizing elements in the trailer.

(d) Degradation of eukaryotic mRNAs is dependent on the action of poly(A) ribonuclease.

(e) Degradation of all mRNAs at the 5' end always involves endoribonucleolytic activity.

(6) **Transfer RNA (tRNA) is functionally involved in the following reactions:**
  (a) transcription
  (b) reverse transcription
  (c) translation
  (d) pre-mRNA splicing
  (e) replication

(7) **The functionality (meaning) of aminoacyl-tRNA is determined by:**
  (a) its amino acid
  (b) its anticodon
  (c) its invariant base regions
  (d) the distance between amino acid and anticodon; the closer the better
  (e) the activity of aminoacyl-tRNA synthetase

(8) **Association of the two ribosomal subunits *in vitro* requires the presence of free Mg^{2+} ions.**
  (a) true   (b) false

Concept questions

(1) Outline the general differences between prokaryotic and eukaryotic mRNAs.

(2) Explain why the name 'messenger' RNA must originate from the study of eukaryotic gene expression rather than prokaryotic gene expression.

(3) Explain why mRNA constitutes only a small proportion (3–5%) of the total cellular RNA.

(4) While the poly(A) tail of eukaryotic mRNAs is not encoded by DNA, it is appropriately added to the 3' end with accuracy. Explain.

(5) Why must mRNA contain untranslated ribonucleotides upstream of the translational start codon?

Summary worksheet: fill in the blanks

Prokaryotic and eukaryotic mRNAs differ fundamentally in their structures (see Parts 3 and 6) and in how they function in the process of translation (Chapters 8 and 9). Nevertheless, the products of .... translation in prokaryotic and eukaryotic cells are identical and they exist as polymerized amino acid molecules, called .... Because of the physical separation of DNA from the site of protein synthesis in .... cells, the foremost property of mRNA is that of an intermediate template for ribosomal amino acid polymerization. This physical separation, however, does not exist in prokaryotic cells and mRNA as a direct product of .... is accessible to ribosomes already in its nascent form. Consequently, the stabilities of prokaryotic and eukaryotic mRNAs have evolved to be quite different with half lives of .... and ...., respectively. While both mRNAs have untranslated regions at the 5' (...........) and 3' (............) ends, prokaryotic mRNA has additional untranslated sequences due to its .... nature. These .... sequences contain the ribosome-binding .... sequences upstream of the translational start codons of individual .... (coding regions, ORFs). The information stored in the .... of mRNA is literally translated into polypeptide sequence by aminoacyl-tRNAs, a process that is facilitated by ribosomes. Each aminoacyl-tRNA interacts with three consecutive nucleotides (= one codon) by recognizing complementarity (Watson–Crick basepairs) to its .... which exclusively determines aminoacyl-tRNA specificity in translation. Therefore, responsibility for the proper conversion of information stored in the .... sequence of DNA into functionality of protein rests in the enzyme that charges a tRNA with the cognate amino acid, which is .... (Chapter 9).

Additional reading

Hentze, M.W. (1991). Determinants and regulation of cytoplasmic mRNA stability in eukaryotic cells. *Biochim. Biophys. Acta* **1090**, 281–292.

Ross, J. (1989). The turnover of messenger RNA. *Scientific American*, August.

Surdej, P., Riedl, A., and Jacobs-Lorena, M. (1994). Regulation of mRNA stability in Development. *Annu. Rev. Genet.* **28**, 263–282.

Chapter 8: Protein synthesis

Multiple-choice questions

(1) What is a ribosome?
- (a) an rDNA-containing organelle in which rRNA synthesis and mRNA translation occur.
- (b) a macromolecular, two-subunit complex comprising more than 50 different proteins and several rRNAs.
- (c) a nuclear protein complex which facilitates intron splicing.
- (d) a structure consisting of three (prokaryotes) or four (eukaryotes) RNAs which facilitates protein synthesis.
- (e) an RNA structure that guides the template selectivity of RNA polymerases I, II and III in eukaryotic cells.

(2) What is the function of initiation factor IF-3?
- (a) if bound to the 40S subunit, it facilitates the association of the 40S and 60S subunits.
- (b) if bound to the 30S subunit, it prevents the association of the 30S and 50S subunits.
- (c) if bound to the 30S subunit, it allows the 16S rRNA of the 30S subunit to interact with the Shine–Dalgarno sequence of the mRNA.
- (d) it directs the initiator tRNA to enter the partial P-site on the 30S subunit bound to mRNA.
- (e) it activates the ribosomal GTPase which enables subunit association upon GTP hydrolysis.

(3) What is the function of eukaryotic initiation factor eIF-3?
- (a) it facilitates formation of the subunit initiation complex (eIF-3, GTP, Met-tRNA, 40S).
- (b) it facilitates binding of the subunit initiation complex (ternary complex, 40S) to bind to the 5' end of mRNA.
- (c) if bound to the 40S subunit, it prevents the association of the 40S and 60S subunits.
- (d) it binds to the cap at the 5' end of mRNA in order to unwind secondary structures.
- (e) it activates the ribosomal GTPase which enables subunit association upon GTP hydrolysis and release of eIF-2.

(4) What is the E-site of the ribosome?
- (a) the site where eukaryotic mRNA is processed.
- (b) an exciting term made up by your instructor.
- (c) the site where tRNA exits the prokaryotic ribosome.
- (d) the site where the endonuclease *Eco*RI restricts the ribosome.
- (e) the site where the electrochemical potential drives translocation.

(5) **Aminoacyl-tRNAs facilitate the translation of how many ribonucleotide triplets?**

(a) 1   (b) 2   (c) 3   (d) 20   (e) 61

(6) **Elongation factor eEF-1α facilitates the entrance of aminoacyl-tRNA into the A-site dependent upon cleavage of a high-energy bond in ATP.**

(a) true   (b) false

(7) **Which of the following statements about the ribosomal peptidyl transferase activity is correct?**

(a) The peptidyl transferase activity resides in the large (50S or 60S) ribosomal subunit.

(b) It facilitates transfer of the C-terminus of the peptide chain from the peptidyl-tRNA to the N-terminus of the aminoacyl-tRNA in the A-site.

(c) It facilitates transfer of the N-terminus of the aminoacyl-tRNA from the A-site to the C-terminus of the peptidyl-tRNA in the P-site by deacylation of aminoacyl-tRNA.

(d) It hydrolyses GTP in order to allow for translocation of the ribosome.

(e) It deacylates peptidyl-tRNA.

(8) **Termination of ribosomal peptide synthesis is achieved by:**

(a) in-frame occurrence of a codon which codes for the C-terminal amino acid.

(b) in-frame occurrence of a codon for which no cognate aminoacyl-tRNA exists.

(c) a too low concentration or the lack of a particular aminoacyl-tRNA.

(d) GTP-dependent action of release factors (RFs) which prevent mismatch binding between termination codons and aminoacyl-tRNAs in the A-site.

(e) the activity of terminal peptidyl transferase, a protein that deacylates peptidyl-tRNA by adding a lysine or arginine residue to the C-terminus of the nascent polypeptide.

(9) **Which of the following compounds is not a product of the initiation reaction?**

(a) GTP + P_i (b) ATP + P_i (c) assembled ribosome

(d) initiation factors   (e) polypeptide

(10) **All three types of RNA have to interact with one another in order to initiate and maintain (elongate) protein synthesis.**

(a) true   (b) false

Concept questions

(1) What is the function of N-formyl-methionyl tRNA?

(2) What are the major differences between a eukaryotic and a prokaryotic ribosome?

(3) Discuss major differences in the initiation of translation in eukaryotes and prokaryotes.

(4) Explain the ribosomal petidyl transferase reaction.

(5) Describe the process of termination of translation in eukaryotes

(6) Describe the synthesis of the eukaryotic 80S ribosomes. Don't forget to indicate where the different parts come from.

Problems

(1) In order to get lots of product for biochemical characterization, you want to express a known (sequenced) gene encoding a monomeric enzyme from the plant *Arabidopsis* in a single-celled microorganism. How will you proceed to ensure that will end up with a product?

(2) You have just been hired by a small biotechnology company to work in their 'antimicrobials' unit. By reviewing the individual steps of translation, you are asked to devise at least six strategies that would result in inhibition of translation.

Summary worksheet: fill in the blanks

Protein synthesis is predominantly accomplished by the translation of .... into amino acid sequence by the .... The ribosome is assembled from two subunits which are complexes consisting of a large highly conserved .... and various specific .... Ribosomal translation of information derived from the DNA by transcription into polypeptide sequence occurs similarly in both prokaryotic and eukaryotic cells: the paradigm of ribosomal protein synthesis lies in the guided base-pairing of mRNA ribonucleotide ...., or codons, with complementary .... of aminoacyl-tRNAs. Thereby the amino acid primary sequence is established following the mRNA codon sequence in 5' to 3' direction. Crucial steps in this process are the finding of the first codon of the information-bearing stretch on the mRNA (the .... or cistron), the supply with needed aminoacyl-tRNAs, the formation of peptide bonds between the .... and the finding of the stop signal which ends the amino acid polymerization process. Ribosome assembly always begins with the facilitated binding of the small subunit to the .... followed by the binding of the initiator tRNA to the partial .... before small and large subunits form the functional ribosome. All these steps occur in a distinctly different manner in prokaryotic and eukaryotic cells. While the eukaryotic facilitator of cytoplasmic translation, the 80S ribosome, cannot read the start position directly and has to scan along the mRNA starting at the cap at the ...., the prokaryotic .... ribosome reads the translational start position directly. This difference already obviates the participation of different .... in eukaryotes and prokaryotes. Eukaryotic mRNA scanning requires binding of factors (eIF-4, 4A and 4B) and the absence of extensive .... structures. The direct 'reading' of the translational start codon by the prokaryotic ribosome is also necessary because of the .... nature of prokaryotic transcripts: every intercistronic (intergenic) sequence must contain a .... in order to attract the ribosome for assembly upstream of the individual cistrons. Specific initiation factors bind to the small subunits to prevent aggregation of the two subunits without binding to translatable message. Other factors are employed to stabilize the functional ribosome once it has assembled and translation was successfully initiated. Study of the ribosomal .... activity has contributed to a completely new chapter of biology with a huge evolutionary dimension: the catalytic activity is suspected to not be of enzymic nature, instead, it might reside in the rRNA. This discovery together with that of autocatalytic pre-mRNA processing (splicing, see Parts 5 and 6) implies that our perceived picture of a cellular world may be just one form of how life is organized at the molecular level.

Additional reading

Condon, C., Squires, C., and Squires, C.L. (1995) Control of rRNA transcription in *Escherichia coli. Microbiol. Reviews* **59**, 623–64.

Harris, E.H., Boynton, J.E., and Gillham, N.W. (1994). Chloroplast ribosomes and protein synthesis. *Microbiol. Reviews* **58**, 700–754.

Jacob, W.F., Santer, M., and Dahlberg, A.E. (1987). A single base change in the Shine–Dalgarno region of 16S . of *E. coli* affects translation of many proteins. *Proc. Natl Acad. Sci. USA* **84**, 4757–4761.

Lake, J. (1981). The ribosomes. *Scientific American*, August.

Noller, H.F. (1993). tRNA–rRNA interactions and peptidyl transferase. *FASEB J.* **7**, 87–89.

Saks, M.E., Sampson, J.R., and Abelson, N.N. (1994). The transfer RNA identity problem: a search for rules. *Science* **263**, 191–197.

Chapter 9: Interpreting the genetic code

Multiple-choice questions

(1) **Which base(s) in the anticodon contribute to the degeneracy (wobbling) in codon reading?**
(a) first   (b) second   (c) third
(d) first and second   (e) second and third

(2) **Which of the four bases, G, A, U, C, lack specificity (unique meaning) at the third codon position?**
(a) G   (b) A   (c) U   (d) C
(e) they always have a unique meaning

(3) **The G•U pairing is responsible for the recognition of GUG by fMet-tRNA$_f$.**
(a) true   (b) false

(4) **'Isoaccepting tRNAs' are:**
(a) multiple tRNAs that recognize synonym mRNA codons (third base degeneracy).
(b) multiple tRNAs that react with the same codon.
(c) multiple tRNAs representing the same amino acid.
(d) multiple tRNAs recognized by the same aminoacyl-tRNA synthetase.
(e) multiple tRNAs that recognize a stop codon thereby causing readthrough.

(5) **Accuracy in the charging of tRNA by aminoacyl-tRNA synthetase is accomplished by:**
(a) chemical proofreading (noncognate aminoacyl-adenylate is hydrolyzed).
(b) kinetic proofreading (noncognate aminoacyl-adenylate dissociates rapidly).
(c) signature sequences in the tRNA D-loop.
(d) the specificity of the amino acid acylation reaction.
(e) a postsynthetic event, i.e, only the correct aminoacyl-tRNA fits into the ribosomal A-site.

(6) **Aminoacyl-tRNA synthetases comprise a family of homologous nucleotide and amino acid-binding enzymes.**
(a) true   (b) false

(7) The number of aminoacyl-tRNA synthetases is:
  (a) determined by the number of existing tRNAs.
  (b) determined by the number of translatable mRNA codons.
  (c) determined by the number of utilized amino acids.
  (d) determined by the individual organismal species.
  (e) not determined by any conservative criterion.

(8) Which of the following contributes to suppressor function?
  (a) use of tRNA mimicking agents such as puromycin.
  (b) creation of a point mutation in the structural gene for tRNA modifying enzymes.
  (c) creation of a mutation in the structural gene for aminoacyl-tRNA synthetase.
  (d) increasing the competition (i.e., release factors, proper tRNAs) efficiency of suppressor tRNA.
  (e) creation of a point mutation in the anticodon region of the gene encoding tRNA.

(9) Codon–anticodon interaction constitutes the weak point in translational accuracy. This is controlled *in vivo* by the following activities:
  (a) selectivity of large subunit proteins for tRNA binding to the A-site.
  (b) differential ribosome velocity dependent on codon-flanking sequences.
  (c) specificity in the interaction between elongation factors and ribosomal A- and P-sites.
  (d) selectivity of small subunit proteins for tRNA binding to the A-site.
  (e) competition of release factors with the tRNAs.

(10) Frameshift suppression can be regarded as a natural translational control mechanism.
  (a) true   (b) false

Concept questions

(1) What is the function of aminoacyl-tRNA synthetase?

(2) Briefly describe the wobble hypothesis.

(3) Why does Ile-tRNA recognize AUU, AUC and AUA but discriminate against AUG?

(4) Why is the presence of strong missense suppressor a big problem for the cell?

(5) Explain the process of kinetic proofreading.

Problems

(1) Your analysis of a DNA contig reveals that the ORF coding for a valuable but labile protein is succeeded by another ORF that would add a stabilizing structure to the C-terminus of the protein. Discuss at least two strategies that will yield the modified protein product.

(2) Two phylogenetically closely related species differ in their chomosomal %[G+C] content, strain A has 55%[G+C] and strain B has 68%[G+C]. Homologous and heterologous transformation/recombination experiments showed that DNA from a lower %[G+C] source transformed easily into the higher %[G+C] organism/genome but not vice versa. Explain.

Summary worksheet: fill in the blanks

The genetic code is universal and encrypts everything necessary to maintain cellular life. The organization of the four .... in groups of three, constituting a codon, gives 64 possible combinations which is considerably more variability than necessary for encoding the 20 .... This has the consequence that almost all amino acids are represented by more than one codon in the range from two to six. Interestingly, this is mostly accomplished by variation in the .... base of the triplet by a lack of discrimination among the pyrimidines (U, C) or purines (A, G). In fact, because methionine and tryptophan are the only amino acids with a unique codon which ends with a G in the third position, C, U and A in the third position never define a particular amino acid. This third base pair .... is enabled by the .... base pairing between the .... codon base and the .... anticodon base, also called 'wobbling' in that 'synonym' triplets code for the same amino acid. While the strart codon of an open reading frame is defined and represented by AUG in eukaryotes and AUG and GUG in prokaryotes, the .... is not translated via complementary base-pairing. Instead, three of the 64 triplets, UAG (amber), UAA (ochre) and UGA, regularly do not match any tRNA anticodon, hence there will only be nonspecific competition for tRNA entrance into the ribosomal .... This competition is regulated by so-called termination factors, proteins that prevent nonspecific binding and facilitate .... of the peptidyl-tRNA in the .... While the transpeptidation could be facilitated by rRNA (Chapter 8), the master players in the information game are enzymes that understand both the amino acid and the ribonucleotide languages, the .... Two families of this enzyme with ten members each control the acylation of tRNA with the correct amino acid. Because this process requires the highest accuracy and fidelity besides that of DNA replication (Part 4), efficient mechanisms of .... and .... have evolved. As potential vehicles of evolution, natural .... (modified anticodons in length or sequence) help to compensate for missense or nonsense mutations. In addition, they provide an efficient tool for differential translational regulation by providing .... suppression.

Additional reading

Cavarelli, J. and Moras, D. (1993). Recognition of tRNAs by aminoacyl-tRNA synthetases. *FASEB J.* **7**, 79–86.

Jukes, T.H. (1977). How many anticodons? *Science* **198**, 319–320.

Mcclain, W.H. (1993). Transfer RNA identity. *FASEB J.* **7**, 72–78.

Schulman, L.H. (1991). Recognition of tRNAs by aminoacyl-tRNA synthetases. *Prog. Nucl. Acid Res. Mol. Biol.* **41**, 23–87.

Chapter 10: Protein localization

Multiple-choice questions

(1) Eukaryotic cells synthesize protein at different locations in the cell, hence they have different kinds of ribosomes. Select the correct one(s) from the following characterizing statements:

   (a) These are the ribosomes bound to the ER (thus called rough ER) and the ones bound to the cytoskeleton (the 'free' ribosomes).

   (b) These are the ribosomes that are single and the ones that are organized in polysomes.

   (c) These are the ribosomes in the cytosol and the ones in the organelles.

   (d) Proteins are synthesized only by ribosomes that are associated with the ER.

   (e) All ribosomes in eukaryotic cells are equal; they do not have different ribosomes.

(2) The general secretory pathway (GSP) describes:

   (a) the mechanisms by which catabolic waste products are exported from the cell.

   (b) the mechanisms by which proteins after ribosomal synthesis reach their functional locations within or outside of the cell.

   (c) the function of the gastro-intestinal tract.

   (d) the mechanisms of nutrient uptake by eukaryotic cells.

   (e) the mechanisms by which a protein is transported across the plasma membrane of a cell.

(3) Chaperones of the Hsp70 family:

   (a) like GroEL, reside in the bacterial cytosol and promote protein folding as well as phage assembly.

   (b) like DnaK, reside in the bacterial cytosol and preserve folding and translocation competence, facilitate protein degradation and reactivate thermally inactivated proteins.

   (c) like SSC1, reside in mitochondria and promote protein translocation into mitochondria and subsequent folding.

   (d) like GrpE, are ATP hydrolases that facilitate polypeptide folding.

   (e) like BiP/Grp78 reside in the ER of mammalian cells and bind unassembled subunits of multisubunit ER proteins.

(4) 'Signal sequence' is:

   (a) a genus/species-specific sequence in 16S (prokaryote) or 18S (eukaryote) rRNA

   (b) the term for the translated leader of an mRNA

(c) the N-terminal amino acid sequence of a preprotein destined for incorporation into or crossing of cellular membranes

(d) the C-terminal amino acid sequence of a protein destined to reside in a specific compartment of the cell

(e) a collective term for molecular tags, e.g. glycosylation, used in protein targeting and sorting

(5) **N-terminal signal sequences (often incorrectly called 'leader sequences') are needed for polypeptide export into cell organelles such as mitochondria or chloroplasts or the incorporation into their membranes. This translocation occurs always post-translationally.**

(a) true   (b) false

(6) **The final destination of organellar proteins is determined by the hierarchy of signal sequences. To export the cytochrome b_2 precursor into the intermembrane space of the mitochondrion, the following is necessary:**

(a) Pre-cytochrome b_2 has one N-terminal signal peptide that is recognized by mitochondrial outer membrane receptor protein. The following interaction with the transport channel protein allows for transport into the intermembrane space where the signal peptide is cleaved.

(b) Pre-cytochrome b_2 has two N-terminal signal sequences. The first one is recognized by mitochondrial outer membrane receptor protein. The following interaction with the transport channel protein, which spans both outer and inner membranes, allows transport into the matrix where the signal peptide is cleaved. This frees the second signal sequence for recognition by and transport through an inner membrane transport channel protein into the intermembrane space where the signal peptide is cleaved yielding cytochrome b_2.

(c) Pre-cytochrome b_2 has one N-terminal signal peptide that is recognized by mitochondrial outer membrane receptor protein. The protein incorporates into the membrane and an ATP hydrolysis-dependent disintegration from the membrane into the intermembrane space is followed by signal peptide cleavage.

(d) Pre-cytochrome b_2 has one C-terminal signal sequence that is recognized by mitochondrial outer membrane receptor protein. The protein crosses the membrane until the C-terminal anchor-stop sequence prevents further translocation. After cleavage of the anchor-stop sequence, the mature peptide is released into the intermembrane space.

(e) Cytochrome b_2 is transported into the mitochondrial intermembrane space by vesicular secretion from the cytoplasm.

(7) **Eukaryotic proteins secreted into the extracellular space or integrated into the plasma membrane are translocated into the lumen of the ER exclusively post-translationally.**

(a) true   (b) false

(8) **Bacteria use both co-translational and post-translational translocation for the export of proteins by the general secretory pathway. This type of pathway is dependent on:**

(a) a C-terminal signal sequence in the protein and an ATP-binding cassette (ABC) in the translocase.

(b) an N-terminal signal sequence and Sec proteins such as SecA, an ATP-hydrolytic peripheral membrane protein, for the localization of the protein at the translocase.

(c) the action of proton motive force for the translocation process.

(d) protein–protein interaction without involvement of a recognized signal sequence.

(e) the action of Hsp70 family chaperones.

(9) **Because the nucleus lacks functional ribosomes, nuclear proteins have to be imported into the nucleus. Which of the following statement(s) provide(s) a correct description?**

(a) All proteins enter the nucleus through nuclear pores by means of active transport.

(b) All proteins enter the nucleus through nuclear pores by passive transport.

(c) Transport into the nucleus is dependent on the protein's molecular mass.

(d) Large proteins (>50 kDa) possess an N-terminal nuclear localization signal.

(e) Translocation of large proteins through the nuclear pore is dependent on ATP hydrolysis.

(10) **What is 'ubiquitin'?**

(a) a subunit of the electron transport chain coenzyme Q, ubiquinone.

(b) a generic name for the most abundant protein in chloroplasts, RubisCo.

(c) a polypeptide involved in marking proteins for degradation.

(d) a polypeptide that modifies a fraction of H2A histone molecules during the cell cycle.

(e) a polypeptide involved in the maintenance of translocation competence of nascent proteins.

Concept questions

(1) Which amino acids are present at the N-termini of unprocessed polypeptide chains in eukaryotes and in prokaryotes?

(2) How does the cell prevent the escape of ER proteins from the ER via transport vesicles to the Golgi apparatus?

(3) Chaperones of the Hsp70 family fulfil various tasks in healthy and stressed cells. List four of them.

(4) Discuss the major difference between co-translational and post-translational translocation.

(5) The text describes an experiment in which the plasma membrane of a eukaryotic cell was permeabilized by digitonin; however, the nuclear membrane stayed intact. What was the aim of the experiment?

Problems

(1) The mRNAs to be translated into polypeptides A and B which assemble to form a soluble enzyme in the intermembrane space are the products of RNA processing in the nucleus and in the mitochondria, respectively. Discuss all steps on the biosynthetic path leading to a functional enzyme. What happens if a mutation cause polypeptide A to bear the C-terminus 'KDEL'? What happens if the cell has a defect in heatshock protein biosynthesis?

(2) Select from the following N-terminal amino acid sequences those suited as signal peptides:

(A) MAADSQTPWLLTFESLKCLHWPQEDQRAGALFGRAKLTR . . .

(B) MMITLRKLMPLAVAVAAGVMSVFAQAMAVETDRAKLE . . .

(C) MPLLNWSRHMVDDTAAKLITVPTRHWLVYATDTLTRDNG . . .

(D) MPEGRRLRRALAIALVPLALVAVTGLLMMAKEQQMGNEI . . .

Summary worksheet: fill in the blanks

Proteins secreted by prokaryotes include a wide range of toxins, degradative enzymes and other factors contributing to stress tolerance, virulence and pathogenicity. This is accomplished by three major, functionally independent .... pathways, called type I, II or III secretion systems. While many proteins are exported across the plasma membrane via the general secretory pathway (Sec pathway, type II), nonproteinaceous compounds, extracellular proteins from Gram-negative bacteria and proteins lacking an N-terminal signal sequence cannot use the Sec pathway. Instead, this is accomplished by two additional major groups

44

of dedicated export systems: the ATP-binding cassette (ABC) transporters (type I) and the type III transporters. ABC transporters operate in both prokaryotes (e.g., histidine uptake and hemolysin export) and eukaryotes (e.g., the P-glycoprotein multidrug resistance pump). Type III secretion appears to occur only in prokaryotes and utilizes components related to the synthesis machinery for type IV pili and flagella to export pathogenicity determinants (e.g. 'harpins', Yops) or enzymes (e.g. pullulanase). This text focuses on the signal sequence-.................. type II general secretory pathway in prokaryotes (SecABDEFY, signal peptidase) as well as the post-translational (into organelles) and co-translational (into the RER) export of .... proteins. Nascent polypeptides destined to leave the compartment need to be kept in a .... conformation which is secured by molecular chaperones of the .... family. Similarly, molecular chaperones of the .... family facilitate proper protein folding after translocation. The N-terminal eukaryotic signal peptide is recognized by a .... (SRP) which facilitates contact between the polypeptide and the translocase via specific interaction with a membrane receptor ('docking') protein. In prokaryotes, the roles of SRP and docking protein are combined in form of the .... membrane protein. 'Delivery' of a translocation-ready polypeptide costs the cell a great deal of energy: GTP hydrolysis is required for dissociation of .... and insertion of the signal sequence into the translocase and ATP hydrolysis is necessary for release of the immature peptide from the .... protein. In addition, protein unfolding as well as initial prevention of polypeptide folding by .... requires additional hydrolysis of ATP. Proteins destined not for export but for incorporation into membranes, initially use the signal recognition/translocation mechanism but terminate translocation by so-called .... sequences which remain in the membrane as 21 amino acid-long .... (MSDs). Dependent on the number of MSDs in a polypeptide, they will incorporate into the membrane following the 'sewing machine principle', thereby creating intracellular and extracellular loop domains of transmembrane proteins. Least understood, so far, is the mechanism of protein export from the cytoplasm into the nucleus (and vice versa), which is assumed to proceed through .... These pores represent molecular mass sieves that do not discriminate against passage of

45

molecules with masses smaller than 50 kDa. Larger nuclear proteins require gated .... which consists of a .... phase and an ATP hydrolysis-dependent .... phase. In addition, nucleoplasmic Ran-GTPase activity is required. Besides translocation, recycling of various overproduced, misdirected or damaged proteins follows protein biosynthetic activities. This is accomplished by .... as well as .... which marks these proteins for degradation and directs them to the proteolytically active ....

Additional reading

Alfano, J.R, and Collmer, A. (1997). The Type III (Hrp) secretion pathway of plant pathogenic bacteria: trafficking harpins, Avr proteins, and death. *J. Bacteriol.* **179**, 5655–5662.

Christie, P.J. (1997). *Agrobacterium tumefaciens* T-complex transport apparatus: a paradigm for a new family of multifunctional transporters in Eubacteria. *J. Bacteriol.* **179**, 3085–3094.

Cline, K. and Henry, R. (1996). Import and routing of nucleus-encoded chloroplast proteins. *Annu. Rev. Cell Dev. Biol.* **12**, 1–26.

Craig, E.A., Gambill, B.D. and Nelson, R.J. (1993). Heat shock proteins: molecular chaperones of protein biogenesis. *Microbiol. Reviews* **57**, 402–414.

Goldberg, A.L. (1995). Functions of the proteasome: The lysis at the end of the tunnel. *Science* **268**, 522–23.

Hartl, F.-U., Hlodan, R. and Langer, T. (1994). Molecular chaperones in protein folding: the art of avoiding sticky situations. *TIBS* **19**, 20–25.

Hicks, G.R. and Raikhel, N.V. (1995). Protein import into the nucleus: an integrated view. *Annu. Rev. Cell Dev. Biol.* **11**, 155–188.

Higgins, C.F. (1992). ABC transporters: from microorganisms to man. *Annu. Rev. Cell Biol.* **8**, 67–113.

Missiakas, D. and Raina, S. (1997). Protein folding in the bacterial periplasm. *J. Bacteriol.* **179**, 2465–2471.

Pfeffer, S.R. (1996). Transport vesicle docking: SNAREs and associates. *Annu. Rev. Cell Dev. Biol.* **12**, 441–461.

Pugsley, A.P. (1993). The complete general secretory pathway in Gram-negative bacteria. *Microbiol. Reviews* **57**, 50–108.

Simonen, M. and Palva, I. (1993). Protein secretion in *Bacillus* species. *Microbiol. Reviews* **57**, 109–137.

Wickner, W.T. (1994). How ATP drives proteins across membranes. *Science* **266**, 1197–1198.

Part 3: Prokaryotic gene expression

Text chapters

Theme

The three chapters of Part 3, 'Prokaryotic gene expression', cover the major characteristics of prokaryotic transcription which encompasses the first stage of gene expression. In Part 2, you already have read about the second stage of gene expression, translation of the coding region(s) of a transcript into the dynamic forces of all living cells, the proteins. This translation is based upon the genetic code and catalyzed by the ribosome, a complex consisting of rRNA associated with various polypeptides. In other words, while transcription selects an area of the genome that contains genetic information at the time when it is needed ('the right information at right time'), translation makes this information available at a needed intensity ('making product at the right time at a physiologically relevant concentration'). This is followed by protein folding and processing ('product available at the right time and location, in a biochemical active form, at a physiologically relevant concentration'). This multi-layered hierarchy of protein biosynthesis provides for the high dynamics of the process and allows for the incredible diversity of life.

While the products of translation, polymers of amino acids held together by peptide bonds, are structurally similar in prokaryotic and eukaryotic cells, the template for translation, the messenger RNA (mRNA), differs with respect to its structure and synthesis. Part 3 addresses the process of RNA synthesis in prokaryotic cells; you will learn about gene expression in eukaryotic cells in Part 6.

One of the reasons for different mRNA synthesis in prokaryotic and eukaryotic cells is that their genomes are organized differently (Part 5). Another reason is that they are dramatically different in their cellular architecture (Part 1). Eukaryotic cells are highly compartmentalized, which physically separates the processes of DNA perpetuation (Part 4) and mRNA synthesis from that of mRNA translation. In contrast, prokaryotic cells lack membrane-bound organelles, hence they represent a single 'bioreactor' in which all three processes and macromolecule synthesis are in a close physical association. Indeed, they occur simultaneously and in a coupled manner. This requires regulatory mechanisms that enables these processes to occur at high speed and prevent the collision between enzymes. Because of limited resources and residence in usually extreme micro-environments, gene expression in prokaryotic cells requires a high plasticity of the process. All this requires:

(1) **An uncomplicated and simultaneous but selective initiation of DNA transcription.** This is achieved by the variable composition of RNA polymerase holoenzyme, Eσ, where the participation of a given sigma subunit ('sigma factor', σ) contributes to recognition of a respective promoter.

(2) **That the transcript contains the genetic information for all proteins that are needed for the completion of an entire biosynthetic pathway or a successful defense against environmental stress.** As initially unraveled by Jacob and Monod, this is accomplished by packaging the genetic information for more than one product into operons that are transcribed from promoters with varying specificity. Prokaryotic transcripts are, thus, polygenic ('polycistronic'). Initiation can proceed as soon as a negative control (*trans*-acting factor) is removed. For multi-enzyme pathways and responses, various operons contribute to a response regulon with a single *trans*-regulatory factor. In other words, the prokaryotic transcription machinery not needed for constitutive (minimal growth-level) gene expression is arrested in 'standby', waiting to be activated. Prokaryotic cells utilize approximately 50% of their genome for constitutive gene expression, hence the other half is reserved for survival of extreme changes in the cell's environment.

(3) **An energy-saving operation and a high efficiency of the process.** This is accomplished by specific termination of the polymerization process and through recycling of the primary transcripts by degradation after a half life of only 10–15 minutes.

Bacteriophages, packaged viral genetic elements that parasitize bacteria, are unable to accomplish self-perpetuation alone, and thus they are operationally dependent on the gene expression machinery of the host. In order to gain control and re-direct host DNA perpetuation and gene expression to perpetuation and expression of the viral genetic information, complex regulatory mechanisms have evolved. These 'phage strategies' will be discussed in Chapter 13.

General reading

Neidhardt, F.C. *et al.* (1989). *Escherichia coli and Salmonella typhimurium—cellular and molecular biology*. Volume 2, American Society for Microbiology, Washington, D.C.

Winnacker, E.-L. (1987). *From Genes to Clones. Introduction to Gene Technology*. VCH Publishers, New York.

Key terms

Use the following key terms as a guide for your study and the composition of the summary at the end of each chapter.

Chapter 11

- abortive initiation
- *cis*-acting element
- coding strand
- consensus sequence
- core enzyme
- DNA
- DNA–RNA hybrid
- elongation
- holoenzyme
- initiation
- intrinsic terminator
- leader
- open binary complex

- *o*pen *r*eading *f*rame
- Pribnow box
- primary transcript
- promoter
- rewinding point
- rho protein
- RNA polymerase
- sigma factor
- SSB
- startpoint (*tsp*)
- stem–loop
- template strand
- terminator

- ternary complex
- topoisomerases I and II
- trailer
- *trans*-acting factor
- transcription
- transcription bubble
- transcription Factor
- translation
- unwinding point
- up- and down- mutations

Chapter 12

- activator
- antisense RNA
- antiterminator
- attenuation
- autogenous regulation
- binary open complex
- cAMP
- CAP
- catabolite repression

- co-repressor
- enhancer
- inducer
- leader transcript
- merodiploid
- modulon
- *o*pen *r*eading *f*rame
- operator
- operon

- plasmid/episome
- polycistronic
- regulon
- repressor
- RNA duplex
- stimulon
- stringent response
- *trans*-acting factors
- tRNA

Chapter 13

- bacteriophage
- constitutive expression
- *cos* site
- Cro protein
- DNA polymerase

- integrase
- λ-repressor
- lysogeny
- lytic cycle
- mRNA

- operator
- pN protein
- pQ protein
- prophage
- transcription

Chapter 11: Transcription

Multiple-choice questions

(1) **Mark ALL correct statements:**

   (a) Transcription is a process that yields, semi-conservatively, two identical strands of DNA.

   (b) DNA-dependent DNA polymerases are multi-subunit enzymes responsible for DNA transcription.

   (c) Bacterial transcripts (mRNAs) are polygenic.

   (d) Sigma factors guide the posttranscriptional modification of eukaryotic hnRNAs to mRNAs.

   (e) Gyrase determines the initiation and termination of transcription by nicking the template strand.

(2) **The DNA-binding protein that initiates the transcription of bacterial genes is called a(n):**

   (a) operator   (b) promoter   (c) repressor

   (d) operon   (e) sigma factor

(3) **The transcriptional startpoint (*tsp*) determines the pyrimidine nucleotide on the template strand where the first RNA–DNA hybrid base pair is established.**

   (a) true   (b) false

(4) **Binding of a given sigma factor depends on the:**

   (a) recognition of the length and spacing of promoter consensus sequences.

   (b) interaction with core enzyme.

   (c) presence of *trans*-acting factors that make up for partial deviation of the promoter from the consensus sequence.

   (d) length of the transcription unit.

   (e) distance of the translational start codon.

(5) **The term 'ternary transcription complex' describes:**

   (a) the complex of sigma factor, core enzyme and double-stranded DNA at the promoter.

   (b) the complex of holoenzyme, TFI and melted DNA duplex.

   (c) the complex of holoenzyme, template DNA and nascent RNA.

   (d) the complex of three holoenzymes at the *tsp*.

   (e) the complex of sigma factor, core enzyme and gyrase.

(6) **The interaction between sigma factor and DNA is best described as follows:**

   (a) The concentration of free and DNA-bound sigma factor is equal; hence, the more sigma factor is synthesized, the higher is the chance for transcriptional initiation.

   (b) Sigma factor is usually DNA-associated and scans along the DNA until it encounters core enzyme at the promoter site. There it binds to DNA independent of core enzyme.

   (c) Sigma factor is usually DNA-associated and scans along the DNA until it encounters a promoter site to which it binds in the presence of core enzyme.

   (d) Sigma factor is the constant part of DNA-dependent RNA polymerase that recognizes the promoter consensus sequence and allows holoenzyme binding.

   (e) Sigma factor joins the ternary complex thereby initiating RNA synthesis.

(7) **Rho-protein-dependent termination occurs more frequently in bacteria than intrinsic termination.**

   (a) true   (b) false

(8) **Read-through of DNA-dependent RNA polymerase can be achieved by:**

   (a) binding of Rho protein to core enzyme.

   (b) binding of antitermination protein to an intrinsic or Rho termination site, thereby masking the termination signal.

   (c) binding of antitermination protein at its utilization site to core enzyme, thereby changing the conformation of and disallowing recognition of the termination signal by core enzyme.

   (d) the binding of the NusA protein to core enzyme.

   (e) polymerase hopping at antiterminator protein–terminator complexes.

(9) **The expression of genes encoding different sigma factors in response to environmental stresses accounts for the regulation of many responses conferring stress tolerance.**

   (a) true   (b) false

(10) **Sigma factor specificity is achieved in that:**

   (a) Sigma factor modifying enzymes (SMEs) enzymically change a generic sigma factor into different specific ones that enable recognition of stress promoters.

   (b) Different genes encode different sigma factors which are able to recognize different promoter consensus sequences.

   (c) Different bacterial species produce different sigma factors which are horizontally exchanged.

    (d) The participation of sigma factor in the initiation complex is dependent on the kind of core enzyme.

    (e) Sigma factor is a nonspecific protein that acts as an accessory factor for every RNA polymerase.

Concept questions

(1) Transcription is an enzyme-facilitated process. Name the substrates, enzymes and products.

(2) Describe the first step of prokaryotic transcriptional initiation in comparison with that of eukaryotic transcription.

(3) Transcription involves the separation of template and coding strands. Explain how the single-stranded DNA is protected during transcription.

(4) Explain in general terms how the accessibility of sigma factor to the promoter is regulated.

(5) What are up- and down- mutations?

(6) Bacterial gyrase mutants are usually not viable; however, gyrase ('ying')/topoisomerase-I ('yang') double mutants can survive. Why?

(7) Explain how the selectivity of sigma factor for specific promoter consensus sequences allows differential gene expression (consider responses to environmental stress).

(8) The *rpoN* gene-encoded sigma factor, σ^{54}, has features different from those of other known prokaryotic sigma factors. Discuss these in comparison with *Escherichia coli* sigma factor σ^{70} (RpoD).

(9) In prokaryotes, there is an equilibrium between loose and tight binding of core enzyme to DNA. Why is this an advantage over an equilibrium between soluble and bound core polymerase?

(10) Review termination of transcription in prokaryotes.

    (a) What are the two major mechanisms of transcriptional termination?

    (b) Describe how translation can regulate transcriptional termination.

    (c) Why is the Rho factor rarely involved in bacterial transcriptional termination?

    (d) How else can the termination of transcription be prevented?

Problems

(1) The biotechnology company providing DNA sequencing services to your laboratory has just returned the following prokaryotic DNA sequence.

```
GAATTCCGGCGGCGGGGTTTTCATTATTATGATCGTTGACATGGGACAAGGGGCTCCTATAATAGGTGC
ATTTGTAGGGGATGTGGCCACATGATAGCGCGTCGAGGCATTCAGGACCGATATTGTCATATTGCCAGC
ATGCTGCAGCGCGCCAACTTAGTTACAACTTATACGATGATTTTAAAAAAAGAATATCAAATAGCCATC
CACCCGAAAGACCACCGCAAGTATCGGTACGATCGTACCAGCACACACCACAGTCGCCAGTCTGTTACC
AAATAATAAAGTTGATAATTAATTCACATCTACCTGGGTAACCCAGGTAGATGTGAATTTTTAGAATTC
```

The sequence insert was cloned into an *Eco*RI restriction site of a multi-copy plasmid. Because you were able to obtain a small peptide expression product you expect to find the following features:

- two *Eco*RI sites
- a promoter
- a transcriptional startpoint
- a transcriptional terminator

With your knowledge from Part 2, you also will identify:

- a ribosome binding site
- a translational start codon
- a translational stop

Also you want to act as the ribosome and provide the translated amino acids in the IUPAC one-letter code for the deduced primary structure.

(2) Polymerases that replicate DNA (DNA-dependent DNA polymerases) also have proofreading function. Explain why RNA polymerases lack this function.

(3) Discuss the importance of the directedness of the polymerization reactions of prokaryotic gene expression. For instance, what would happen if the ribosome were to read its template, mRNA, from the 3' to the 5' end?

Summary worksheet: fill in the blanks

Prokaryotic DNA-dependent RNA polymerase interacts directly with DNA. One of its subunits, the ...., recognizes and binds to a highly conserved, ... element: the .... (of transcription). ...., made up of four subunits ($2\times$ α, $1\times$ β and $1\times$ β'), and sigma (σ) factor assemble at the promoter to the ...., Eσ. Both β and β' subunits constitute the catalytic center and, not surprisingly, share a high degree of similarity with the large subunits of eukaryotic RNA polymerases (see Chapter 28). The α subunit dimer appears to be involved in holoenzyme assembly as well as in promoter recognition. RNA polymerase recognizes promoters with different .... due to the specificity of different .... These consensus sequences are two conserved blocks of 5 to 8 nucleotides located approximately 10 and 35 nucleotides upstream of the mRNA start site, the .... (*tsp*) located at '+1' in the DNA. The promoter consensus regions are, therefore, called .... and .... regions. While the .... is implicated primarily in recognition and binding of RNA polymerase, the ...., also known as ...., is thought to be the location where the DNA duplex is melted. Once bound, holoenzyme and promoter form a closed binary complex which changes into an .... upon melting of the duplex. This 'tight binding' is usually irreversible. Mutagenesis of the promoter region will alter the binding of .... to .... and, consequently, initiation efficiency. Dependent on whether transcription is enhanced or reduced, these mutations are called .... or .... mutations.

RNA polymerase will facilitate the placement of ribonucleotides at the .... which are complementary to the template DNA. The formation of phosphodiester bonds between the ribonucleotides creates the .... which together with RNA polymerase is also called .... The nascent RNA can grow up to .... long before the transcription complex needs to move forward (5' to 3') on the DNA. If the chain is not released prematurely (called ....), the chain will grow as the holoenzyme moves along the ....; transcription has been successfully initiated. At this stage, the .... has been created and .... will be released.

Elongation is accomplished by .... RNA polymerase moves along the DNA reading the .... of DNA in a .... to ... direction thereby polymerizing a nascent RNA in the ... to ... direction which is identical to the .... of DNA. Because DNA exists conformationally in a double-helical form, the DNA has to be unwound and rewound which creates the so-called .... This is accomplished by .... at the .... and .... points. During elongation, the melted single-stranded DNA in the transcription bubble is protected by .... In contrast, the nascent RNA emerges in a ..... to ...... direction from .... as a .... Elongation will occur as long as RNA polymerase does not encounter a .... signal. This signal can be either an .... structure or the .... protein.

54

The first occurs in the nascent RNA and stalls the enzyme. If the DNA template at this (growing) point contains a stretch of ...., the .... interaction is extremely weak and both .... dissociate. This leaves the RNA polymerase without a free .... for further extension, and the transcription is terminated. Transcription and translation are tightly coupled in prokaryotes, hence a normal situation would be that a translating .... chases a transcribing .... Another way of transcriptional termination is accomplished by the .... that has to bind to the .... of the .... and sneak up in order to interact with RNA polymerase. Because of the tight coupling of transcription and translation ....- dependent termination of prokaryotic transcription is rare.

Additional reading

Benson, A.K. and Haldenwang, W.G. (1993). *Bacillus subtilis* σ^B is regulated by a binding protein (RsbW) that blocks its association with core RNA polymerase. *Proc. Natl Acad. Sci. USA* **90**, 2330–2334.

Gralla, J.D. (1991). Transcription control—lessons from an *E. coli* promoter data base. *Cell* **66**, 415–418.

Henkin, T.M. (1996). Control of transcription termination in prokaryotes. *Annu. Rev. Genet.* **30**, 35–57.

Kustu, S., Santero, E., Keener, J. Popham, D. and Weiss, D. (1989). Expression of σ^{54} (*ntrA*)-dependent genes is probably united by a common mechanism. *Microbiol. Rev.* **53**, 367–376.

Loewen, P.C. and Hengge-Aronis, R. (1994). The role of the sigma factor σ^S (*katF*) in bacterial global regulation. *Annu. Rev. Microbiol.* **48**, 53–80.

McClure, W.R. (1985). Mechanisms and control of transcription initiation in prokaryotes. *Annu. Rev. Biochem.* **54**, 171–204.

Merrick, M. (1993). In a class of its own—the RNA polymerase sigma factor σ^{54} (σ^N). *Molec. Microbiol.* **10**, 903–909.

Morett, E., and Segovia, L. (1993). The σ^{54} bacterial enhancer-binding protein family: mechanism of action and phylogenetic relationship of their functional domains. *J. Bacteriol.* **175**, 6067–6074.

Pérez-Martîn, J., Rojo, F., and de Lorenzo, V. (1994). Promoters responsive to DNA bending: a common theme in prokaryotic gene expression. *Microbiol. Reviews* **58**, 268–290.

Pribnow, D. (1975). Bacteriophage T7 early promoters: nucleotide sequences of two RNA polymerase binding sites. *J. Molec. Biol.* **99**, 419–443.

Straus, D.A., Walter, W.A. and Gross, C.A. (1987). The heat shock response of *E. coli* is regulated by changes in the concentration of σ^{32}. *Nature (London)* **329**, 348–351.

Chapter 12: The operon

Multiple-choice questions

(1) A series of genes that are transcriptionally regulated in *trans* by DNA-binding proteins are called a(n):
   (a) transcriptor   (b) repressor   (c) promoter
   (d) operon   (e) oppressor

(2) Most regulatory elements acting at the level of bacterial transcription are negative controls.
   (a) true   (b) false

(3) DNA-dependent RNA polymerase will transcribe the *lac* operon constitutively from the *E. coli* chromosome if a missense mutation-containing *lacI* gene is expressed from a plasmid. This shows conclusively that:
   (a) Repression of *lac* operon transcription is dependent on repressor concentration.
   (b) The functional *lac* repressor is an oligomer and the mutation was *trans*-dominant.
   (c) Homologous recombination has occurred.
   (d) The LacI protein can also up-regulate transcription by binding to RNA polymerase core enzyme.
   (e) The antisense strand of the plasmid insert encoded a *lac* inducer that derepresses the *lac* operon.

(4) Transcription of the six pre-rRNA gene copies (rDNA operons) of *Escherichia coli* is accomplished from extremely strong promoters that initiate about 50% of all transcription activity in exponentially growing cells. This high promoter strength has been related to the high binding affinity of the otherwise poorly characterized α subunit dimer to the rDNA promoters.
   (a) true   (b) false

(5) Regulation of polygenic transcription in prokaryotes can be at the level of:
   (a) RNA-polymerase holoenzyme recruitment to DNA (sigma factor specificity for promoters).
   (b) enhancer binding protein-induced isomerization of the binary complex (DNA:Eσ).
   (c) repressor and activator protein binding near the promoter element.
   (d) termination and antitermination of transcription.
   (e) all of the above.

(6) **Once an operon has been transcribed, differential expression must be regulated at the level of translation. This is often accomplished by protective binding of proteins to the mRNA at or near translational initiation sites (Shine–Dalgarno sequence).**

 (a) true   (b) false

(7) **Most DNA-binding proteins interact in *trans* with DNA elements by:**

 (a) covalent binding to the phosphodiester moieties of the nucleic acid backbone.

 (b) fitting a recognition α-helix into the minor groove of one half-site of the DNA element.

 (c) inserting their phosphorylation sites into the major groove of the DNA element.

 (d) fitting a recognition α-helix into the major groove of one half-site of the DNA element.

 (e) all of the above.

(8) **The stringent response is a prokaryotic cell's response to adverse growth conditions. It involves:**

 (a) activity of the (p)ppGpp synthetase which is associated with the A-site of the ribosome.

 (b) derepression of the genes that encode most hexose-utilizing enzymes (i.e., glucose effect).

 (c) activity of aminoacyl-tRNA synthetase which loads tRNA with the dearly needed amino acids.

 (d) the LexA and RecA proteins which up-regulate recombinational repair during starvation.

 (e) none of the above.

(9) **A helix-turn-helix motif describes:**

 (a) the basic structural unit of the DNA double helix.

 (b) a recurring structural element in many prokaryotic DNA binding proteins, i.e., repressors.

 (c) the organization of promoter elements.

 (d) the structure of DNA in an open binary complex.

 (e) the interaction between an enhancer-binding protein and promoter-bound RNA polymerase.

(10) **The following are basic properties of a prokaryotic operon. The operon**

 (a) usually consists of multiple genes that are organized between a promoter and a terminator.

 (b) the operon is usually under autogenous transcriptional control.

 (c) Member genes are preceded by identical promoters.

 (d) Member genes are coordinately regulated.

 (e) The operon contains exclusively genes that encode proteins involved in stress tolerance.

Concept questions

(1) One product of the *lac* operon expression is β-galactosidase, which catalyses the hydrolysis of lactose into galactose and glucose. The galactoside moiety of lactose has inducer function and de-represses the *lac* operon upon interaction with the *lac* repressor, the LacI protein. Explain the reasons for using X-gal (5-bromo-4-chloro-3-indolyl-β-D-galactoside) and IPTG (isopropyl-β-D-thiogalactoside) successfully in a colorimetric assay for β-galactosidase (including 'blue–white screening').

(2) Two *lacI⁻* mutants of *Escherichia coli* express β-galactosidase constitutively. Merodiploids were constructed by transformation with plasmids that carry complete functional *lac* operon inserts. One of the two merodiploid *lacI⁻* mutant strains reverted to the wild-type phenotype with inducible *lac* gene expression. The other merodiploid *lacI⁻* mutant strain maintained constitutive *lac* gene expression. Explain.

(3) Discuss the phenomenon of catabolite repression which is the regulatory mechanism behind diauxic growth. Involve scenarios of glucose, lactose, galactose and arabinose catabolism.

(4) Draw a diagram showing the characteristic features of a regulon (e.g., *trp*, *oxyR*)

Problems

(1) You are interested in the molecular characterization of temperature-induced exopolysaccharide (EPS) synthesis. An increase in EPS requires the up-regulation of various enzymes and it appears that this up-regulation occurs at the level of gene expression rather than enzyme activity. Discuss two different strategies that will allow you to estimate or determine the length of the temperature-inducible EPS operon.

(2) Recombinant DNA work, i.e., the cloning of the foreign DNA, can become impossible if any one of the expression products is toxic to the host cells. Discuss methods that theoretically will prevent expression while allowing DNA replication. For example, consider (a) a scenario in which the DNA insert lacks an indigenous promoter, (b) one in which the sequence of the transcribed message is partially known from cDNA sequencing and (c) yet another one which involves restriction mapping.

(3) The OxyR protein is a transcriptional regulator that is central to the hydrogen peroxide stimulon. It binds at sites near promoters that regulate transcription of operons involved in the prevention and repair of damage resulting from exposure to hydrogen peroxide. Thus, OxyR is an activator exerting positive control on transcription. Besides this regulatory function, the protein also functions as an oxidative stress sensor. This information should be sufficient to discuss properties of OxyR biosynthesis and function.

(4) A universal prokaryotic mechanism of response to environmental cues and stresses is the utilization of two-component regulatory systems. A receptor/transmitter molecule (usually histidine protein kinase; Hpk) autophosphorylates upon signal reception and interacts with a response regulator. Discuss this interaction and the effector function in bacteria for the case of (a) nutrient chemotaxis, (b) toxin resistance and (c) nitrogen metabolism.

(5) A three-subunit enzyme with a bottleneck function in a single substrate-dependent catabolic pathway needs to replenish its catalytic site due to the fact that the product is toxic and needs to be metabolized immediately to avoid malfunction. Propose a map of the enzyme-encoding operon with emphasis on the transcriptional controls.

(6) You are asked to develop an experimental scheme that allows you to prove that the product of the *lacI* gene, which precedes the promoter/operator region of the *lac* operon, binds the *lacO* element. By employing the same technique you should also demonstrate that binding of RNA polymerase and *lac* repressor (LacI) are mutually exclusive. Further, you are asked to show evidence for the regulation of the glucose effect (catabolite repression).

Summary worksheet: fill in the blanks

The control of gene expression in single-celled prokaryotes serves mainly to empower the cell with quick and flexible responses to environmental signals and cues. This is in contrast to multicellular eukaryotes in which the regulation of gene expression follows genetic programs in processes of differentiation and development. Prokaryotic gene expression responds to abruptly fluctuating physical and chemical parameters including excess or gradual depletion of essential nutrients (nutrient stress or starvation), exposure to noxious chemicals (chemical stress) as well as competition for surface sites (physical stress). Against intuitive expectations, well nurtured, rapidly growing prokaryotic cells are highly susceptible to environmental stress, while starved cells have a high stress tolerance. Remember that stationary phase, starved cells are physiologically 'sedate', maintain only one copy of their genome and have a reduced rRNA content with the consequence of an overall low rate of protein synthesis. Naturally, this provides fewer opportunities for interference of stresses with metabolic and gene expression activities. The magic that prokaryotic transcriptional and translational regulation can work becomes evident in that stressed or resting bacteria can switch to high levels of gene expression and metabolic

59

activities within just minutes; even entire populations are able to respond synchronously (e.g., motility waves of swarming bacteria such as *Proteus* or *Myxococcus*). These well-defined and specific responses to environmental stresses can be explained only by the existence and operation of multigene and global control systems that are organized along strict hierarchical principles.

The smallest expression unit is the ...., which represents a cluster of one or more contiguous genes that are transcribed coordinately from a single promoter. Transcription is terminated downstream from the last operon member gene by mechanisms such as intrinsic termination or the action of the .... protein. The product, a .... (multigenic) mRNA, is usually translated into polypeptides that contribute to a common functional goal (a biochemical pathway; synthesis and secretion of a defense enzyme). Hence, which and how much of a particular mRNA is going to be produced is determined by the process of .... initiation. This represents a crucial control point. This 'up-front' regulation approach is accomplished by processes of RNA polymerase recruitment and isomerization which involve the binding of *trans*-acting factors to *cis*-acting elements as well as the interaction between proteins bound to DNA rather far apart. These transcriptionally regulatory *cis*-elements are the ...., ... and .... sequences. The first is selected by the specificity of ...., which allows initiation of transcription of selected operons based upon sigma factor availability. The second provides an opportunity for DNA-binding proteins to bind and, consequently, positively or negatively affect RNA-polymerase binding to the promoter. These DNA-binding proteins are .... and .... (which exert negative and positive control of transcription, respectively). Both can be involved in the induction and repression of transcription which is initiated by .... and ...., respectively. By interaction with an active repressor, an inducer weakens the interaction between repressor and .... and thus de-represses transcription. In contrast, a co-repressor enables stable binding interaction between repressor and operator; initiation of transcription is prevented. Similarly, activators can be activated and deactivated resulting in induction and repression, respectively.

More complex and hierarchically higher regulatory units can be defined along genotypic or phenotypic lines. Genotypically, if multiple, unlinked operons are controlled by the same protein repressor, they constitute a .... Member operons usually contribute to a more complex (but single) metabolic pathway or function and the .... is named after the gene encoding the repressor (e.g., OxyR regulon). In a ...., member operons are controlled by a pleiotropic regulatory protein and contribute to multiple pathways (e.g., cAMP–CAP modulon). In addition, member operons may still be governed by individual regulatory proteins. Phenotypically, a set of independently co-induced (derepressed or activated) or co-repressed operons that respond to a single environmental signal constitute a .... Stimulon member operons may belong to various regulons and modulons, and thus do not share a regulatory protein (e.g., hydrogen peroxide stimulon). This means that membership of an operon in different regulatory units creates overlap in the response where different stimuli/stresses cause related response phenotypes. This is particularly evident in the coordinated regulation in operon networks such as the multigene systems responsible for cell division or sporulation. Other global control systems such as the heat shock or SOS response are often called 'adaptive response systems' and they entail multi-pathway responses to a specific, well-defined stress.

Interestingly, transcriptional activity can also be regulated by the ribosome. This is caused by the close physical and temporal association of transcription and translation in prokaryotic cells. In a process called ...., an RNA-polymerase-chasing ribosome determines the timing of intrinsic termination by being close to core enzyme (antiterminator can not form, terminator formation yields termination) or by lagging behind (antitermination allows core enzyme to continue transcription). Slowing of the ribosome stems from insufficient availability of selected aminoacyl-tRNAs: it is logical that this mechanism regulates the expression of genes that encode enzymes involved in biosynthesis of the very amino acids that the cell needs to charge the respective tRNAs. The ribosome facilitates the response to amino acid starvation also posttranscriptionally. Association with the RelA protein near the A-site allows the synthesis of (p)ppGpp upon encounter of uncharged tRNAs. While the ribosome becomes idle, (p)ppGpp effects various changes in the cell's overall transcriptional activity that are collectively called the .... Most remarkably, it down-regulates rRNA synthesis. This in turn will cause autogenous down-regulation of r-protein operon transcription. Another

61

mechanism of posttranscriptional regulation is the signal-stimulated synthesis of antisense RNA. Concomitant .... formation will protect the double-stranded stretch of the mRNA from translation.

Additional reading

Becker, A., Rüberg, S. Küster, H. Roxlau, A.A. Keller, M. Ivashina, T. Cheng, H.-P. Walker, G.C. and Pühler, A. (1997). The 32-kilobase *exp* gene cluster of *Rhizobium meliloti* directing the biosynthesis of galactoglucan: genetic organization and properties of the encoded gene products. *J. Bacteriol.* **179**, 1375–1384.

Christman, M.F., Morgan, R.W., Jacobson, F.S. and Ames B.N. (1985). Positive control of a regulon for defenses against oxidative stress and some heat-shock proteins in *Salmonella typhimurium*. *Cell* **41**, 753–762.

Czarniecki, D., Noel, R.J. Jr. and Reznikoff, W.S. (1997). The −45 region of the *Escherichia coli lac* promoter: CAP-dependent and CAP-independent transcription. *J. Bacteriol.* **179**, 423–429.

Gottesman, S. (1984). Bacterial regulation: global regulatory networks. *Annu. Rev. Genet.* **18**, 415–442.

Kustu, S., North, A.K. and Weiss, D.S. (1991). Prokaryotic transcriptional enhancers and enhancer-binding proteins. *Trends Biochem. Sci.* **16**, 397–402.

Magasanik, B. (1989). Gene regulation from sites near and far. *New Biol.* **1**, 247–251.

Mager, W.H. and De Kruijff, A.J.J. (1995). Stress-induced transcriptional activation. *Microbiol. Reviews* **59**, 506–531.

Pabo, C.O. and Sauer, R.T. (1992). Transcription factors: Structural families and principles of DNA recognition. *Annu. Rev. Biochem.* **61**, 1053–1095.

Platt, T. (1986). Transcription termination and the regulation of gene expression. *Annu. Rev. Biochem.* **55**, 339–372.

Reznikoff, W.S. (1992). Catabolite gene activator protein activation of *lac* transcription. *J. Bacteriol.* **174**, 655–658.

Saier, M.H. and Ramseier, T.M. (1996). The catabolite repressor/activator (Cra) protein of enteric bacteria. *J. Bacteriol.* **178**, 3411–3417.

Stevens, A.M., and Greenberg, E.P. (1997). Quorum sensing in *Vibrio fisheri*: Essential elements for activation of the luminescence genes. *J. Bacteriol.* **179**, 557–562.

Storz, G., Tartaglia, L.A. and Ames, B.N. (1990). Transcriptional regulator of oxidative stress-inducible genes: direct activation by oxidation. *Science* **248**, 189–194.

Chapter 13: Phage strategies

Multiple-choice questions

(1) **Lytic (bacterio)phages may have the following genetic elements:**
    (a) ssRNA only   (b) dsRNA only   (c) ssDNA only
    (d) dsDNA only   (e) all of the above

(2) **Prophage, i.e. phage genome integrated into a bacterial genome, is selfish DNA because:**
    (a) it always propagates together with the genome of the lysogen.
    (b) it replicates independently of the genome of the lysogen.
    (c) it provides immunity against infection by further virions.
    (d) (a) and (c)
    (e) (b) and (c)

(3) **In order to recombine the phage attachment site (*attP*) with the attachment site in the bacterial chromosome (*attB*), the DNA of lambda phage (infecting *E. coli*) becomes circular after infection by annealing of the terminal *cos* sites.**
    (a) true   (b) false

(4) **A virus has a genome consisting of single-stranded minus-strand RNA. Which scheme correctly portrays how the viral mRNA is formed?**
    (a) the viral single-stranded minus-strand RNA serves directly as mRNA.
    (b) double-stranded RNA is made and the new plus strand serves as mRNA.
    (c) single-stranded plus-strand DNA is made which is transcribed into mRNA.
    (d) the genome of single-stranded minus-strand RNA viruses only codes for viral tRNAs.
    (e) none of the above

(5) **Lysogeny is the state in the 'life' cycle of a double-stranded DNA virus when the viral genetic element (prophage) is integrated into the genome of a host cell (lysogen).**
    (a) true   (b) false

(6) **The transition from lysogeny to the lytic cycle of temperate phages is characterized by one of the following processes:**
    (a) repression of the autogenously regulated expression of the *cI* gene by Cro protein.
    (b) repression of the expression of the *cro* gene by lambda repressor protein.

(c) induction of prophage replication by transition factor (TF)-mediated recruitment of DNA polymerase

(d) transition sigma factor-mediated induction of transcription from the P_R promoter.

(e) none of the above

(7) **A critical influence over the switch between lysogeny and the lytic cycle is the stability of the cII protein:**
(a) true   (b) false ·

(8) **The λ repressor protein is a DNA-binding protein that belongs to the group of:**
(a) helix-loop-helix proteins
(b) helix-turn-helix proteins
(c) zinc finger proteins
(d) leucine zipper proteins
(e) DNA demethylases

(9) **Enzymes for lambda phage DNA synthesis are made during the late phase of the lytic cycle:**
(a) true   (b) false

(10) **Expression of late phase genes of lambda prophage is enabled by:**
(a) polymerase recruitment: sigma factor, σ^{RE}, initiates transcription from promoter P_{RE}.
(b) antitermination: factor N allows read-through of transcription from promoter P_R.
(c) repression: Cro protein represses transcription from promoter P_R.
(d) antitermination: factor Q allows read-through of transcription from promoter P_R'.
(e) none of the above.

Concept questions

(1) What would happen without the timely regulation of phage transcription when all genes of a virulent phage are expressed?

(2) Why is a lambda phage-created lysogen usually 'immune' against re-infection by another lambda phage?

Problems

(1) Make predictions regarding the phenotypes of lambda phage-infected bacteria conferred by the mutations described in the following. Explain your reasoning:

(a) a mutation creating a protease-resistant λ cII protein.

(b) a mutation of λ O_{R2} that will prevent protein binding.

(c) a mutation that inactivates the λ N gene.

(d) a mutation in the gene encoding the λ cI protein.

(2) The commonly employed test for the screening of carcinogenicity of compounds is the Ames test which looks for revertant frequencies of auxotrophic bacteria. Design an experiment using lambda phage and *E. coli* that could be useful for carcinogenicity testing.

Summary worksheet: fill in the blanks

Viruses can cause lytic, persistent or latent infections of host cells that either destroy the invaded host immediately or they can place the cell's genetic and metabolic machinery under immediate (persistent) or future (latency) control of the virus for a longer period of time. Bacterial viruses or .... attack solely bacteria, hence they prey on single cells. Most phages can establish a latent infection in their hosts, called ...., which allows propagation and survival of both bacterium and phage. The lysogenic nature of phage–bacterial interactions is host specific, which means that a given virus may be always lytic (virulent) in one bacterium while it may be lysogenic and capable of conversion to the lytic cycle (temperate) in another bacterial species. This requires that the viral genome is protected from destruction by host restriction systems during the latent phase and that it provides information allowing for differential regulation of viral replication and expression. This is needed independent of whether or not the lytic phase was preceded by lysogeny because phages are usually complex structures and assembly is solely controlled by the specificity and affinity of the viral parts for one another.

The best studied temperate phages are the so-called 'T-even' phages and lambda phage. Upon infection (injection of its double-stranded DNA), lambda phage DNA (λDNA) is circularized and .... facilitates integration of .... into the bacterial genome in a single reciprocal crossover event. Whether lambda remains lysogenic or goes into the lytic phase is ultimately determined by the concentration of two proteins, the .... and the .... (or ... protein), which compete for binding at the same transcriptional .... located between oppositely directed promoters, P_{RM} ('repressor maintenance'; leftward) and P_R ('rightward'). P_{RM} initiates transcription of the .... gene, cI, whereas the rightward transcription initiated from P_R will yield the .... transcript. .... binds cooperatively and with high affinity for operator regions O_{R1} and O_{R2} which will allow positive autogenous regulation of cI gene expression from P_{RM}. In addition, if .... can bind to the O_{R1} and O_{R2} regions, it will prevent binding of .... to the O_{R3} and O_{R2} operator regions as well as the initiation of cro gene transcription by .... from P_R (which overlaps with O_{R1}). Conversely, Cro protein binds with higher affinity to operator regions O_{R3} and O_{R2} and prevents the binding of .... to the O_{R1} and O_{R2} operator regions that is required for expression from P_{RM} (which overlaps with O_{R3}). In essence, a high cI to Cro protein ratio will inhibit transcription from P_R and thus maintain .... while the degradation of .... will allow for conversion into .... The integrity of .... is maintained by functional cII and cIII proteins, whose synthesis is dependent on successful transcription from the 'rightward' and 'leftward' promoters, P_R and P_L.

The complex regulation of the lytic cycle is mostly accomplished by antitermination of transcription (read-through of DNA polymerase) which establishes two characteristic phases. The early phase accounts for expression of Cro protein and antiterminator protein, pN, which is needed for the 'delayed early' transcription of genes encoding factors essential for recombination and replication of λDNA as well as antiterminator protein, ... The latter will facilitate initiation and cause read-through of transcription from late phase promoter $P_{R'}$. These late .... will be then translated into head and tail parts of the virus.

Additional reading

Birge, E.A. (1994). *Bacterial and bacteriophage genetics*. Springer Verlag, New York.

Black, L.W. (1989). DNA packaging in dsDNA phages. *Annu. Rev. Microbiol.* **43**, 267–292.

Daniels, D., Sanger, F. and Coulson, A.R. (1983). Features of bacteriophage lambda: analysis of the complete nucleotide sequence. *Cold Spring Harbor Symposium Quant. Biol.* **47**, 10009–1024.

Echols, H. (1986). Bacteriophage λ development: temporal switches and the choice of lysis or lysogeny. *Trends in Genetics* **2**, 26–30.

Fields, B.N. and Knipe, D.M. (eds) (1995). Field's Virology. Raven Press, New York.

Lwoff, A. (1953). *Lysogeny. Bact. Rev.* **17**, 269–337.

Part 4: Perpetuation of DNA

Text chapters

Theme

Part 4 is concerned with the synthesis, maintenance/repair and expansion of DNA, hence it addresses mechanisms which contribute to both conservation and alteration of the genetic information in a cell. The modus operandi of a particular mechanism depends greatly on DNA topology and size and—not surprisingly—there are significant differences between prokaryotic (usually circular DNA) and eukaryotic (linear DNA) genomes. Despite these differences, it is not difficult to produce a general definition of the structural unit of replication, or 'replicon'. Any DNA that is not separated from an origin by a termination site is part of the same replicon. The problems arise because (i) DNA polymerases require a free 3'-end and a template for DNA synthesis (problem: initiation of replication at the end of a linear chromosome), (ii) a large linear chromosome contains many bidirectional replicons (problems: termination by replicon fusion; ensuring that every replicon is active only once) and (iii) the bacterial cell cycle can be much shorter than the time required for replication and segregation (problem: regulation of initiation at the origin). You will be amazed, again, that the nucleic acid structure in form of an antiparallel duplex of complementary nucleotide polymers inherently bears simple solutions also to the mentioned problems of perpetuation. Key to this is the fact that DNA replicates semiconservatively, hence 'new' is always paired with 'old' in the process. And because 'new' stays with 'old' after the process, the cell even has the opportunity to control the product of replication for accuracy and—if a mistake is recognized—repair it according to the 'old' sequence. To this end, the cell has to maintain a large toolkit of DNA modifying enzymes such as nucleases, polymerases, ligases, helicases, topoisomerases and methylases, some of which become available or active in response to regulatory signals.

Interestingly enough, there is no successful replication without the function of DNA-dependent RNA polymerases (transcriptases) and even RNA-dependent DNA polymerases (reverse transcriptases). While RNA polymerases ('primases') provide oligonucleotide primers for replication at the origins and the priming sites of Okazaki fragment synthesis at the the lagging DNA strand in both prokaryotic and eukaryotic cells, reverse transcriptases are needed for initiation of replication (5' to 3') at the 3'-

ends of the linear eukaryotic chromosomes, the telomeres. This will prevent shortening and, hence, the loss of genetic information.

How we view the consequences of all the powerful DNA modifying tools depends largely on the definition of their original purpose; are recombinational DNA expansion (gene duplication and conversion), transposition and error-prone DNA repair 'negative' in that they disrupt the fidelity of DNA perpetuation? Or are they part of the original evolutionary 'design' controlled by the semiconservative nature of replication and a wealth of 'conservation-oriented' repair mechanisms which limit the extent of change to usually incremental levels? No matter what answer your personal perspective will yield, the material presented in Part 4 underlines again that the simple 'makeup' of the DNA molecule allows both simple and fairly complex modifications to be made without compromising its function: to be the blueprint of life.

General reading

Lodish, H. *et al.* (1995). *Molecular Biology of the Cell*, 3rd edition. W. H. Freeman & Co., New York.

Stryer, L. (1995). *Biochemistry*. 4th edition. W. H. Freeman & Co., New York.

Key terms

Use the following key terms as a guide for your study and the composition of the summary at the end of each chapter.

Chapter 14

❏ amplification	❏ helix	❏ semiconservative
❏ cDNA	❏ mutation	❏ single-stranded
❏ chromatin	❏ recombination	❏ telomerase
❏ chromosomal	❏ replication	❏ template
❏ cytidine	❏ replicon	❏ terminal protein
❏ extrachromosomal	❏ rolling circle replication	❏ theta replication
❏ fork meeting point		

Chapter 15

❏ DNA ligase	❏ helicase	❏ replicase
❏ DNA duplex melting	❏ hemimethylated DNA	❏ replisome
❏ DNA polymerase I, DNA polymerase III	❏ Okasaki fragment	❏ restriction
❏ DnaA, DnaB, DnaC	❏ primase	❏ supercoiling
❏ exonuclease	❏ primosome	❏ termination site
	❏ repair	❏ unwinding

Chapter 16
- daughter strands
- DNA ligase
- DNA polymerase I
- endonucleases
- foreign DNA
- hemimethylation
- LexA protein
- methylase
- mismatch
- parental strand
- RecA protein
- recombinational repair
- SOS response
- template
- UvrABC

Chapter 17
- chiasma
- chromosomes
- exonuclease
- gene conversion
- homologous sequences
- intasome
- linking number
- mutation
- meiosis
- patch recombinant
- RecA protein
- reciprocal recombination
- RuvAB proteins
- single-strand exchange
- site-directed recombination
- splice recombinant
- supercoiling
- template
- topoisomerase

Chapter 18
- *Ac* element
- antibiotic resistance
- autonomous elements
- bacteriophage Mu
- breakage–fusion–bridge cycle
- cointegrate
- composite transposon
- conservative transposition
- controlling elements
- direct repeat
- *Ds* element
- hybrid dysgenesis
- insertion sequence
- inverted repeat
- nonautonomous elements
- nonreplicative transposition
- P element
- replicative transposition
- resolvase
- *Spm* element
- Tn10
- TnA
- transposase
- transposon

Chapter 19
- Alu element
- *copia* element
- delta
- *env*
- FB element
- first jump
- *gag*
- helper virus
- integrase
- L1
- LINES
- LTR
- nonviral superfamily
- *onc*
- oncogenic virus
- *pol*
- provirus
- replication deficient virus
- retrotransposon
- retrovirus
- reverse transcriptase
- second jump
- SINES
- transducing virus
- transforming virus
- tRNA primer
- Ty element
- viral superfamily

Chapter 14: The replicon

Multiple-choice questions

(1) **A replicon is:**
   (a) the segment of DNA that after replication is segregated during cell division.
   (b) the segment of DNA to be replicated and the enzymes and proteins needed in the process.
   (c) any autonomously replicated DNA sequence that is not separated in *cis* from an origin by a terminus.
   (d) the product of any given replication mechanisms (e.g., single circle).
   (e) the segment of DNA between the origin and the replication fork.

(2) **The following features are common to all (prokaryotic, eukaryotic, viral) replication origins:**
   (a) Origins are unique DNA segments that contain multiple short repeated sequences.
   (b) Origins are palindromic sequences that form stable secondary structures.
   (c) Multimeric DNA-binding proteins specifically recognize these short repeat sequences.
   (d) The origin-flanking sequences are A•T-rich to allow melting of the DNA duplex.
   (e) The origin-flanking sequences are G•C-rich to stabilize the initiation complex.

(3) **Rolling circle replication:**
   (a) is the dominant mechanism of replication in bacteria.
   (b) allows for amplification of the original replicon.
   (c) always generates double-stranded circular copies of the original replicon.
   (d) is a common mechanism of bacteriophage DNA replication in bacteria.
   (e) is autoregulatory in that the gene for the nicking protein resides in the replicon.

(4) **Replication of DNA by DNA-dependent DNA polymerase requires the presence of a free 3'-OH that acts as a primer. This can be accomplished, for example, by synthesis of an RNA primer, DNA self-priming or a terminal protein linked through the phosphodiester covalent bond to a nucleotide.**
   (a) true   (b) false

(5) **Mammalian mitochondria and plant chloroplast genomes replicate by D-loop replication. Which of the following statements describe the process accurately?**

    (a) Both strands are replicated from *oriD*, which is a unique secondary structure recognized by DNA polymerase recruiting proteins.

    (b) Replication of both strands is initiated simultaneously from two independent origins.

    (c) Replication of both strands is initiated subsequently from two independent origins.

    (d) Initiation of replication is facilitated by the presence of one or two (strand) displacement loops.

    (e) Completion of replication of one strand is delayed (by the *ter* locus) until both replication processes are in synchrony.

(6) **The eukaryotic unit of segregation coincides with the unit of replication. This ensures that replication is truly completed before the cell cycle advances from S/G$_2$ into M phase.**

    (a) true   (b) false

(7) **Eukaryotic replicons have the following properties. They:**

    (a) are much shorter than prokaryotic replicons because of the presence of termination sequences.

    (b) are much longer than prokaryotic replicons because of the larger genome.

    (c) are usually bidirectional and able to fuse.

    (d) all 'fire' at once to ensure that the entire chromosome is replicated during S-phase.

    (e) do not all 'fire' at once: only approximately 15% are active at any given moment.

(8) **Plasmid survival in a bacterial population is ensured by (mark ALL correct statements):**

    (a) dependence of division upon a minimum or maximum number of origins of replication per cell.

    (b) the presence of 'addiction systems' (killer and antidote).

    (c) natural competence and horizontal transfer (conjugation).

    (d) the operation of dedicated segregation systems.

    (e) the plasmid residence of genes encoding proteins with vital function.

(9) **Initiation of plasmid replication from a ColE1 origin in *Escherichia coli* is controlled by countertranscriptional regulation in which an antisense RNA forms a stable duplex with the primer RNA.**

    (a) true   (b) false

Concept questions

(1) Describe the Meselson and Stahl experiment and explain its importance for our understanding of inheritance.

(2) Describe a system in which replication of the two strands of a circular duplex DNA is initiated from independent and separated origins (e.g., D-loop replication).

(3) Discuss three ways in which linear replicons can be established at the end of linear chromosomes.

(4) Explain the process of episome conjugation (F-plasmid, Hfr).

(5) Discuss mechanisms that ensure survival and incompatibility of plasmids in a bacterial cell.

Problems

(1) How is it possible that some bacteria complete binary fission in less than the time needed for replication of the bacterial genome? Explain why *Escherichia coli* can afford a multiforked chromosome or even up to four decatenated chromosome copies under optimal nutritional conditions, considering that the chromosome is 'normally' under single copy control.

(2) In order to create a stable mutation, you want to transfer a mutagenized gene cloned from a genomic library of *Pseudomonas aeruginosa* from the recombinant cloning host (*Escherichia coli*) back into the wild-type organism. Devise a strategy by utilizing your knowledge about conjugation and plasmid incompatibility.

Summary worksheet: fill in the blanks

The entire genome of a cell, ... as well as .... DNA, has to be replicated in order to maintain its hereditary information. In addition, the regulation of this process must ensure that the replication of the DNA occurs only .... per cell cycle which is completed with the generation of ...... daughter cells. Surprisingly, this is accomplished by a very simple but effective mechanism called .... replication. This means that there is no free or ....-independent synthesis of DNA; new always pairs with old (as shown by the Meselson and Stahl experiment). In other words, genetic/hereditary information is either maintained (by ....), acquired (by ....) or created by alteration of existing information (................ and ....) but not created .... Mechanistically, the process of replication is fairly complicated; you have to remember that the nucleic acid duplex is wound up into a .... which is organized at various further levels into .... structures. All these structures have to be undone in an orderly manner in order to 'melt' the two DNA strands and replicate, and the process must also allow the fast and faithful reorganization of the two replica duplex structures.

The structural unit of replication is the .... which is defined by a site of replication initiation, the ...., and its termination. By recognizing the ......-acting nature of an origin, one can say that any DNA that is not separated from an origin by a termination site belongs to the same replicon as the origin. The replicon extends on the DNA molecule either unidirectionally or bidirectionally; most replicons are established simultaneously from overlapping origins in the duplex, hence they are .... In the case of prokaryotic circular DNA such as bacterial and plasmid DNAs, the most common mechanism of replication is the so-called .... replication in which the two replication forks move away and towards one another at the same time, hence the original and the replica do not go through .... intermediates. Termination is accomplished by termination sites which are located downstream of the .... Replication of circular DNA with a single-stranded intermediate is the .... replication (example: circularized phage DNA). This requires a single-strand break at the initiation site and allows .... of the original replicon into linear, ..., single- or multi-copy replicas which need protection from degradation (by SSB). All eukaryotic, some prokaryotic (spirochetes) and some plasmid DNAs are linear. Because of this linear nature and because of the much larger size of the eukaryotic genome, replication of eukaryotic DNA involves many .... of which only about 15% are active at any given time during ..-phase. While this constitutes a regulatory problem in terms of ensuring that all replicons fire only once, the real challenge comes with the initiation and termination of individual replicons. Termination is usually achieved by the .... of bidirectional replicons, a process that is still not completely understood. The complicated mechanism of priming at the 3' ends of the linear DNAs is better understood. In phage DNA, polymerization is primed by a .... (covalently linked to the template 5' end) which

73

provides an extendable .... nucleotide (cytidylate) 3'-OH group. In eukaryotic chromosomes, this is accomplished by an RNA-dependent ...... polymerase (called ....) that generates .... from an RNA primer, thus it is a reverse transcriptase (see Chapter 26).

The presence of multiple replicons in eukaryotic linear DNA constitutes another difference between prokaryotic and eukaryotic cells in that the unit of replication does not coincide with the unit of segregation. Thus, the mechanisms reponsible for limiting the activity of a replicon in eukaryotic linear DNA to only once between .... must be different from those in prokaryotic circular DNA. This is addressed in detail in Chapters 26 and 15.

Additional Reading

Addinall, S.G., Bi, E., and Luktenhaus, J. (1996). FtsZ ring formation in *fts* mutants. *J. Bacteriol.* **178**, 3877–3884.

de Boer, P.A.J., Cook, W.R., and Rothfield, L.I. (1990). Bacterial cell division. *Annu. Rev. Genetics* **24**, 249–274.

Frost, L.S., Ippen-Ihler, K., and Skurray, R.A. (1994). Analysis of the sequence and gene products of the transfer region of the F sex factor. *Microbiol. Reviews* **58**, 145–161.

Jaffé, A., Vinella, D., and D'Ari, R. (1997). The *Escherichia coli* histone-like protein HU affects DNA initiation, chromosome partitioning via MukB, and cell division via MinCDE. *J. Bacteriol.* **179**, 3494–3499.

Kuempel, P.L., Pelletier, A.J., and Hill, T.M. (1989). Tus and the terminators: the arrest of replication in prokaryotes. *Cell* **59**, 581–583.

Kues, U., and Stahl, U. (1989). Replication of plasmids in Gram-negative bacteria. *Microbiol. Reviews* **53**, 491–516.

Novick, R.P. (1987). Plasmid incompatibility. *Microbiol. Reviews* **51**, 381–395.

Zakian, V.A. (1996). Structure, function, and replication of *Saccharomyces cerevisiae* telomeres. *Annu. Rev. Genetics* **30**, 141–172.

Zyskind, J.W., and Smith, D.W. (1986). The bacterial origin of replication, *oriC*. *Cell* **46**, 489–490.

Chapter 15: DNA replication

Multiple-choice questions

(1) Mark ALL correct statements:

  (a) Transcription is a process that yields, semi-conservatively, two identical double strands of DNA.

  (b) DNA-dependent DNA polymerases are multi-subunit enzymes responsible for DNA replication.

  (c) Bacterial transcripts (mRNAs) are polygenic.

  (d) Sigma factors guide the posttranscriptional modification of eukaryotic hnRNAs to mRNAs.

  (e) Gyrase (topoisomerase II) determines the initiation and termination of replication by nicking the template strand.

(2) DNA polymerization requires the presence of template and of a free 3'-OH end. This end may be generated by:

  (a) synthesis of an RNA primer at the origin or at Okasaki fragment start points (3'-GTC).

  (b) nicking one strand of duplex DNA followed by strand displacement.

  (c) random Watson–Crick base-pairing of free deoxyribonucleotides with template.

  (d) by loop formation at the 3' end (self-priming).

  (e) binding of a terminal nucleotide-carrying protein to the 3' end of template.

(3) Replication of DNA employs both DNA polymerases and RNA polymerases.

  (a) true   (b) false

(4) Dependent on a given origin, a 'primosome' consists of:

  (a) an oligomeric enzyme that interacts with DnaG primase at initiation sites.

  (b) a single-strand binding protein that prevents DNA degradation.

  (c) DnaB helicase and accessory proteins such as DnaC, DnaT or PriA.

  (d) DnaB, single-strand binding protein, DnaC, DnaT, PriA proteins and DnaG primase.

  (e) DnaB helicase, DnaG primase and DNA polymerase III.

(5) **Bacterial DNA polymerase III is asymmetrical, containing six subunit complexes ($\gamma,\delta,\delta',\tau,\psi$ and χ) in addition to the catalytic core. These subunits are implicated in the following functions:**

    (a) γ recruits (in a process involving ATP-hydrolysis) the β subunit dimer to the DNA at the primer-template site where it forms the 'clamp'.

    (b) δ causes the catalytic core (α, ε, θ) to dimerize and it interacts with DnaB helicase.

    (c) The γ-facilitated binding of the β subunit dimer to the DNA recruits the catalytic core of polymerase III to the lagging strand.

    (d) After the τ-facilitated core dimer has bound to the leading strand, γ facilitates binding of PolIII to the lagging strand.

    (e) After contact with DnaB helicase, τ mediates the interaction between DnaG and the five asymmetrical subunit complexes ($\gamma,\delta,\delta',\tau,\psi$ and χ) causing a priming event.

(6) **While DNA replication of both strands occurs simultaneously and is facilitated by one enzyme complex, DNA polymerase III, eukaryotic replication employs three independently acting DNA polymerases, one copy of Polα (for initiation) and two copies of Polδ (DNA polymerization, fill-in after RNA primer removal by MF1).**

    (a) true   (b) false

(7) **Which enzyme removes the RNA primer and fills in deoxyribonucleotides in prokaryotic replicons?**

    (a) DNA polymerase III

    (b) DNA polymerase II

    (c) DNA polymerase I

    (d) exonuclease MF1

    (e) DNA ligase

(8) **Establishment of the forks during initiation of replication entails the following steps:**

    (a) DnaA protein cooperatively binds to 9 bp repeats causing the oriC region to wrap around the DnaA cluster. This allows DnaA to interact with the A•T-rich 13 bp repeats resulting in the melting of the duplex and the formation of two forks.

    (b) DnaA protein binds to the A•T-rich 13 bp repeats and recruits the DnaB/DnaC complex to the origin. This allows DnaB to melt the DNA duplex and create the forks.

    (c) Topoisomerase II binds to the A•T-rich 13 bp repeats and begins to relax duplex supercoiling. This will melt the DNA duplex eventually and create the forks.

    (d) Single-strand binding protein binds to the A•T-rich 13 bp repeats and initiates strand displacement facilitated by the DnaA protein.

This will create a replication 'eye' flanked by two forks, where primase can bind.

(e) DnaA, DnaB and DnaC form the initiation complex which binds to the 9 bp repeats near the origin. DnaB helicase unwinds the DNA until the 13 bp repeat region becomes part of the open circle. Additional bending of this structure by the HU protein allows primase binding at the replication forks.

(9) Origins of the *oriC* type contain 11 copies of a sequence that is recognized by Dam methylase indicating that the methylation state of the origin plays a role in the initiation process. Select all correct statements:

(a) Because initiation of replication occurs only at nonmethylated DNA, the action of Dam methylase is a negative regulatory mechanism that prevents premature replication.

(b) Hemimethylated origins, generated as the result of replication, cannot be initiated again before methylase remethylates the origin. Hence, prevention of remethylation prevents premature replication.

(c) Because initiation of replication occurs only at fully methylated DNA, the action of demethylase prevents initiation of hemi- and nonmethylated origins.

(d) Hemimethylated origins, generated as the result of replication, cannot be initiated again because the DnaA protein has a lesser affinity for hemimethylated DNA than a cytoplasmic membrane protein. Membrane association can be overcome only by remethylation.

(e) DnaA protein has ATP-dependent methylase activity and remethylates hemimethylated origins after replication. This process takes ~13 min because DnaA synthesis is also regulated by the methylation state of the *dnaA* promoter.

(10) Initiation of phage replication from *oriλ* is controlled by two phage proteins, O and P, which are analogs of the DnaA and DnaC proteins in *Escherichia coli*. Based on this comparison, the O-protein represents a helicase while the P-protein regulates helicase and primase binding.

(a) true   (b) false

Concept questions

(1) What processes occur before polymerization of the new DNA strands by DNA polymerase III gets under way?

(2) How is this initiation process affected by the methylation state of the DNA?

(3) How is the replication of the lagging strand accomplished?

(4) What phenotype occurs if base pair errors are generated and how can they be repaired?

(5) Discuss one striking difference between DNA replication initiated at *oriC* or φX type origins.

Problems

(1) Eukaryotic replication has to be regulated so that each of the many replicons is activated only once. This could be achieved by negative (repression) or positive (rate-limiting factor) regulation. Devise an experiment that demonstrates that the regulation is, indeed, carried out by a protein (complex).

(2) *Escherichia coli* OH157 can grow incredibly quickly. Under normal (poor environmental) conditions, these bacteria divide after 60 to 70 min indicating that replication and cell division are successive processes, one only being initiated after completion of the other. If given the opportunity to live in a juicy and warm, half-cooked beefburger, the doubling time is close to 20 minutes which is twice as fast as the standard time required for the process of replication. Explain. Give a detailed description of the DNA replication process in prokaryotes (e.g., the structural and functional characteristics and a sequence of events). How is it possible that the replication machinery involved in multiple genome copy replication does not collide?

Summary worksheet: fill in the blanks

While the replicon is the .... unit of replication and, hence, clearly defined, the replication process involves various enzyme complexes that form dynamically as they are needed. Four processes are to be considered: (i) initiation of replication at the .... by a proteinaceous initiation complex; (ii) repeated initiation of replication at the lagging strand by the ....; (iii) elongation (unwinding or the parental strands and synthesis of the daughter strands) by the .... and (iv) termination of the process at .... or by replicon .... The key step of initiation is the .... and .... of the DNA duplex at the origin which is accomplished by the DnaA and DnaB/DnaC proteins, respectively. While the .... and .... proteins are specific to the initiation process at the origin, the .... helicase is also part of the .... and .... complexes. In fact, it constitutes the minimum primosome because in ....-type replicons DnaB is sufficient to interact with and locate the DnaG .... to the initiation sites of .... fragment synthesis on the .... strand. In other types

78

of replicons such as phage φX, additional priming proteins (e.g., PriABC) assist DnaG .... with the recognition. .... assembly and disassembly also happens in linear eukaryotic replicons, hence it is a periodical process that generally occurs multiple times at the lagging strand near the replication fork in an active replicon.

Elongation of the primer, or daughter strand synthesis along the parental template is facilitated by DNA-dependent DNA polymerases and accessory enzymes. First, a reduction in DNA .... is necessary which is accomplished by topoisomerases. Second, DnaB .... (as a member of the primosome) unwinds the helix, melts the DNA in the fork and facilitates synthesis by interacting with .... While there is similarity between some subunits of eukaryotic and prokaryotic polymerases, the priming and organization of the elongation complex are significantly different in prokaryotes and eukaryotes. The DNA replicase in bacteria is an asymmetric dimer of ...., where the asymmetry stems from additional subunits that allow simultaneous synthesis of leading and lagging strands. This is remarkable because lagging strand synthesis occurs in a direction .... to the movement of the fork. Prokaryotic .... exercises its .... activity (RNA primer removal) and it extends the previous (free 3' end) to the new (5' end) Okazaki fragment before the phosphodiester bond is introduced by DNA .... In contrast, RNA primer removal and replacement by DNA synthesis are carried out by two different enzymes, the .... MF1 and the DNA .... that catalyzes nucleotide polymerization of the leading strand and the Okazaki fragments. All this contributes to the approximately ten times .... speed of prokaryotic compared with eukaryotic replication. Last, another important feature of replication is that the newly synthesized daughter strands initially lack any methylation, hence both duplexes are .... Recognition of the DNA methylation state contributes to the regulation of various mechanisms that follow DNA synthesis such as the .... of unfaithful replication and mutations, reinitiation of the next replication cycle and segregation (in prokaryotes). In addition, it renders the DNA vulnerable to the cell's own .... system.

Additional reading

Coverley, D. and Laskey, R.A. (1994). Regulation of eukaryotic DNA replication. *Annu. Rev. Biochem.* **63**, 745–776.

Firshein, W. (1989). Role of the DNA/membrane complex in prokaryotic DNA replication. *Annu. Rev. Microbiol.* **43**, 89–120.

Kelman, Z. and O'Donnell, M. (1995). DNA polymerase III holoenzyme: structure and function of a chromosomal replicating machine. *Annu. Rev. Biochem.* **64**, 171–200.

Kogoma, T. and Maldonado, R.R. (1997). DNA polymerase I in constitutive stable DNA replication in *Escherichia coli*. *J. Bacteriol.* **179**, 2109–2115.

Lake, J.A. (1988). Origin of the eukaryotic nucleus determined by rate-invariant analysis of rRNA sequences. *Nature* (London) **331**, 184–188.

Lohman, T.M. and Ferrari, M.E. (1994). *Escherichia coli* single-stranded DNA-binding protein: multiple DNA-binding modes and cooperatives. *Annu. Rev. Biochem.* **63**, 527–70.

Maas, R., Wang, C., and Maas, W.K. (1997). Interactions of the RepA1 protein with its replicon targets: two opposing roles in control of plasmid replication. *J. Bacteriol.* **179**, 3823–3827.

McHenry, C.S. (1988). DNA polymerase III holoenzyme of *Escherichia coli*. *Annu. Rev. Biochem.* **57**, 519–550.

Nossal, N.G. (1983). Prokaryotic DNA replication systems. *Annu. Rev. Biochem.* **53**, 581–615.

Radman, M. and Wagner, R. (1988). The high fidelity of DNA duplication. *Scientific American*, August.

Chapter 16: Restriction and repair

Multiple-choice questions

(1) **The most common modification of DNA is methylation. In prokaryotes, this serves to:**
   (a) identify damaged DNA for repair.
   (b) distinguish the strands after replication and allow/disallow continued replication.
   (c) identify methylated foreign DNA for recombinational incorporation into the genome.
   (d) protect its own DNA but not foreign DNA from restriction by endonucleases.
   (e) identify the transcriptional startpoint and recruit RNA polymerase.

(2) **The recognition and cleavage sites of type I restriction enzymes are usually separated more than 1000 bp. If a type I enzyme recognizes hemimethylated DNA, it will methylate, whereas if it encounters nonmethylated DNA, it will cleave energized by hydrolysis of ATP.**
   (a) true   (b) false

(3) **Type II restriction enzymes:**
   (a) have endonuclease and methylase activity and often recognize palindromic sequences.
   (b) have endonuclease activity only while methylase activity resides in a different enzyme.
   (c) are 'restricted' in that they recognize only nonmethylated characteristic nucleotide sequences.
   (d) have exonuclease and methylase activity.
   (e) have exonuclease activity only while methylase activity resides in a different enzyme.

(4) **Type III restriction enzymes:**
   (a) consist of two subunits that recognize only hemimethylated sites. The location of the methylation site with respect to the restriction site (upstream or downstream) decides whether the DNA is methylated or restricted.
   (b) have mutually exclusive methylation and restriction activity dependent on which of the subunits recognizes the site: MS facilitates methylation, R facilitates restriction.
   (c) consist of two subunits that facilitate recognition and methylation or restriction near the recognition site.
   (d) are the key players in mismatch repair because binding to DNA is based upon structural distortion rather than sequence error recognition.

81

(e) are light-dependent enzymes that are crucial to 'photoreactivation' of pyrimidine dimers because they are able to cleave the covalent bonds between crosslinked adjacent thymine bases.

(5) **Short-patch repair is an inducible function of the bacterial cell that must be induced by damage. In contrast, long-patch repair is a constitutive function and entails the replacement of stretches of injured DNA extending from 1500 up to 9000 bp.**

(a) true   (b) false

(6) **Single base changes constitute one form of injury in DNA. They:**

(a) affect transcription but not replication in that an ATG start codon could be modified.

(b) affect the sequence but not the overall structure of DNA.

(c) will continue to cause structural problems over many replication cycles.

(d) can arise from mismatch synthesis or enzymic DNA modification such as deamination.

(e) can arise from UV irradiation (e.g. pyrimidine dimer) or adduct formation (e.g. alkylation).

(7) **Mismatch repair is based on the recognition of mismatches that arise during replication.**

(a) The UvrABC system recognizes and excises mismatches to be repaired by DNA polymerase I-facilitated introduction of the correct nucleotide.

(b) If recognition occurs before hemimethylated DNA by remethylated, repair can be directed with bias toward the wild-type sequence (Dam methylase, MutH, MutSL).

(c) Mismatches are commonly repaired by single-strand exchange which depends on the ability of the RecA protein to retrieve the sequence of the normal copy.

(d) Mismatch repair describes DNA modifying activities such as dealkylation or reamination, but never the replacement of damaged nucleotides.

(e) Mismatch repair is accomplished by repair functions normally repressed by the LexA protein (SOS response).

(8) **Repair of DNA in eukaryotes is less important than in prokaryotes because of the diploid character of somatic cells. So far, retrieval systems (post-replication repair), excision repair and recombinational repair systems have been identified. DNA injury in transcriptionally active genes is preferentially repaired.**

(a) true   (b) false

Concept questions

(1) Discuss why most endonucleases are called 'restriction' enzymes.

(2) Discuss why the methylation state of DNA can be utilized for the regulation of replication and repair of DNA.

(3) Describe the individual steps of the Uvr excision repair system in *Escherichia coli*.

(4) Discuss how the direction of mismatch repair can be regulated (mutant to wild type or wild type to mutant).

(5) Describe how the RecA protein regulates the SOS response.

Problems

(1) Your investigation reveals that the bacterium you are studying has three nearly identical copies of an operon that codes for the subunits of an essential enzyme. No third basepair degeneracy (wobbling) is detectable which suggests that the near identity (99%) is the result of a DNA modifying activity. Discuss which of the modification and repair systems covered here might be the best candidate for maintenance of this DNA identity.

(2) Based upon your knowledge of restriction and repair systems in bacteria, devise a simple experiment to test your collection of bacterial strains for presence of prophages.

Summary worksheet: fill in the blanks

Because they are haploid, bacteria are under tremendeous stress to defend the integrity of their DNA against contamination by .... DNA as well as against loss of crucial hereditary information caused by faulty replication or mutations. To this end, bacteria contain specific .... that recognize characteristic sequences as well as the state of methylation to distinguish between ...... and .... DNA. These restriction systems have to recognize foreign DNAs before they are replicated because host .... will remethylate any DNA in a manner specific to the cell that presents the appropriate recognition sequences.

The methylation state of the DNA after replication also provides crucial information for systems involved in the repair of injured DNA. The .... allows the distinction between the .... and the .... strands and, hence, correction of a mismatch or injury by using the parental strand as the .... Excision repair, which can be very short, short or long, involves the activity of the .... protein complex (which has recognition and endonuclease activity), ...

(which has exonuclease activity and synthesis of DNA replacement) and .... (which seals nicks). Many mutator loci have been identified whose products confer the ability to carry out .... repair. These ...... gene products are able to specifically replace individual mismatched nucleotides such as T from G•T and C•T mismatches.

Repair after replication and remethylation may involve using the healthy daughter duplex for repair of the injured or faulty daughter duplex. In addition to endo- and exonuclease activities, this repair involves the .... protein which facilitates single strand invasion and strand displacement in recombination. For this reason, retrieval systems are also known as .... repair systems. If the RecA protein is activated for repair, it will trigger the induction of expression of genes that are usually repressed by the .... protein. Many of these products have repair functions and the response that can occur as fast as minutes after the DNA injury is known as the ...... response.

Additional reading

Britt, A.B. (1996). DNA damage and repair in plants. *Annu. Rev. Plant Physiol. Plant Mol. Biol.* **47**, 75–100.

Demple, B. and Harrison, L. (1994). Repair of oxidative damage to DNA: enzymology and biology. *Annu. Rev. Biochem.* **63**, 915–948.

Drake, J.W. (1991). Spontaneous mutation. *Annu Rev. Genetics* **25**, 125–146.

Echols, H., and Goodman, F. (1990). Mutation induced by DNA damage: a many protein affair. *Mutation Research* **236**, 301–311.

Hoeijmakers, H.J.H. (1993). Nucleotide excision repair I: from *E. coli* to yeast. *Trends in Genetics* **9**: 173–177.

Lindahl, T. (1990). Repair of intrinsic DNA lesions. *Mutation Research* **238**, 305–311.

Messer, W. and Noyer-Weidner, M. (1988). Timing and targeting: the biological functions of Dam methylation in *E. coli*. *Cell* **54**, 735–737.

Modrich, P. (1991). Mechanisms and biological effects of mismatch repair. *Annu. Rev. Genetics* **25**, 229–253.

Murli, S. and Walker, G.C. (1993). SOS mutagenesis. *Curr. Opin. Genet. Dev.* **3**, 719–725.

Prakash, S., Sung, P, and Prakash, L. (1993). DNA repair genes and proteins of *Saccharomyces cerevisiae*. *Annu. Rev. Genetics* **27**, 33–70.

Sancar, A. (1995). DNA repair in humans. *Annu. Rev. Genetics* **29**, 69–105.

Van Houten, B. (1990). Nucleotide excision repair in *Escherichia coli*. *Microbiol. Reviews* **54**, 18–51.

Wilson, G.G. (1988). Type II restriction-modification systems. *Trends in Genetics* **4**, 314–319.

Chapter 17: Recombination

Multiple-choice questions

(1) **Recombination that:**
  (a) occurs between sequences that contain homologous genes is called homologous recombination.
  (b) occurs between precisely corresponding sequences (so that no single base pair is added or lost) is called homologous recombination.
  (c) occurs between sequences that contain segments of highly similar sequences is called homologous recombination.
  (d) occurs between one duplex of any homologous sequences is called homologous recombination.
  (e) occurs between specific pairs of sequences and is facilitated by enzymes that only act on the particular target sequences is called site-specific recombination.

(2) **Recombination can be induced by single-strand and by double-strand breaks:**
  (a) A single-strand break is required for the assimilation of a homologous single strand.
  (b) A single-strand break is required for reciprocal single-strand exchange between two DNA duplexes.
  (c) A double-strand break initiates D-loop-like strand displacement of the donor duplex and the following single-strand DNA synthesis, nicking and ligation creates two patch recombinant duplexes.
  (d) In order to avoid the loss of genetic information, repair of a double-strand break requires the retrieval of sequence from another, homologous duplex.
  (e) A double-strand break is required for initiation of RecBCD-facilitated recombination.

(3) **The reciprocal exchange of DNA from the two duplexes during recombination creates a recombinant joint where a segment of an individual single strand from one duplex pairs with the complementary strand of the other duplex. This formation of heteroduplex DNA requires the presence of a nick in one strand of each duplex and these strands must be complementary. Dependent on whether a second nick occurs in the same or the complementary strand in each duplex, patch recombinant (heteroduplex segment) or splice recombinant DNAs are created (make a drawing!).**
  (a) true   (b) false

(4) **Branch migration of a Holliday junction is necessary to extend the repair/recombination segment into homoduplex regions of the interacting double strands. The migration is achieved by:**

(a) the ATP-hydrolysis-dependent activity of the RecA protein.

(b) the ATP-hydrolytic action of the RuvAB helicase complex.

(c) the activity of the RecBCD protein complex.

(d) the tension in the DNA molecule caused by coiling.

(e) random base pairing between the invading strand and the recipient strand.

(5) **Many bacteria have almost evenly distributed hotspots in their genomes that stimulate recombination. These hotspots, called *chi* in *Escherichia coli*, are:**

(a) locations of frequent double-strand breaks which in turn induce recombination.

(b) locations of frequent single-strand breaks which lead to single-strand assimilation.

(c) sites at which double-strand break-activated RecBCD complex cuts a free 3'-OH end.

(d) *cis*-acting elements that allow the generation of single strands with free 3'-OH end.

(e) the DNA-binding sites for the RecA protein from which it scans along the DNA until a break is encountered.

(6) **While the exchange of duplex DNA is common in eukaryotes (e.g., double crossover during meiosis), recombinational exchange of duplex DNA in the haploid prokaryotes is rare.**

(a) true   (b) false

(7) **The topological state of DNA is determined by the linking number which is a function of the twist of DNA around an axis and the writhe of the double helix. Combined, this is known as supercoiling and is affected by the following enzymes:**

(a) Type I DNA topoisomerases which nick a single strand and relax highly positively supercoiled DNA.

(b) Type I DNA topoisomerases which nick a single strand and relax highly negatively supercoiled DNA.

(c) Type II DNA topoisomerases which generally relax both negatively and positively supercoiled DNA by nicking both strands.

(d) selected Type II DNA topoisomerases such as gyrase that can introduce negative supercoils into relaxed DNA circles.

(e) selected Type II DNA topoisomerases such as gyrase that can relax only positive but not negative supercoils.

(8) **Lambda phage integrates its double-stranded DNA into the bacterial genome where it becomes an integral part of the bacterial chromosome as a 'prophage'. Phage DNA integration is accomplished by:**

    (a) restriction of the bacterial genome with *Lam*RI followed by integration of the phage DNA facilitated by integrase.

    (b) site-specific (reciprocal) recombination between the phage *attP* and the bacterial *attB* sequences (staggered cleavage at the attB and attP sites provides complementary single-stranded ends which can form a Holliday junction).

    (c) by single-strand assembly (linear DNA is incorporated into the bacterial genome).

    (d) homologous recombination (precise assimilation of identical DNA).

    (e) intasome-facilitated interaction between specific phage and host DNA sequences.

Concept questions

(1) Explain the generation and resolution of a Holliday (junction) structure during recombination.

(2) How is it possible to rescue DNA information after suffering a double strand break?

(3) Explain the mechanism(s) that allow the movement of the branchpoint.

(4) What is gene conversion? Consider postmeiotic segregation in eukaryotes.

(5) How is it possible to introduce a negative supercoil into circular DNA?

Problems

(1) Gene conversion can be studied using multiply marked diploid yeast. While normal reciprocal crossover (homologous recombination) during meiosis will segregate the allelic markers in a 2:2 ratio, nonreciprocal crossover or gene conversion modifies this segregation ratio postmeiotically into 3:1. Discuss the given sequences by locating the site of heteroduplex formation (A+a' and A'+a) and discussing segregation if restriction is used as an allelic marker.

    A   5'-GCGCATTACTCGAATTCTACGACGTCTAT-3'

    A'   3'-CGCGTAATGAGCTTAAGATGCTGCAGATA-5'

    a   5'-GCGCATTACTCGATTACTACGACGTCTAT-3'

    a'   3'-CGCGTAATGAGCTAATGATGCTGCAGATA-5'

(2)   During recombination, DNA duplex molecules can interact (join) at either homologous or nonhomologous sites thereby forming a Holliday structure. In one of the two micrographs available for evaluation, the lengths of the 'arms' protruding from the Holliday junction are equal. In contrast, all 'arms' in the other structure are of different length. Use a sketch to discuss which micrograph showed a product of homologous recombination. Indicate the nick-sites that will produce patch recombinants and splice recombinants.

Summary worksheet: fill in the blanks

While .... allow small changes in the genome that may or may not manifest themselves in the phenotype, ... generates significant changes and can be viewed as the motor of evolution. Recombination involves the exchange of large segments of DNA within or between ...., thus it occurs usually between two DNA duplexes. However, DNA duplexes can also assimilate single-stranded DNA which is mediated by the .... protein. The reciprocal exchange of .... (sharing a common ancestor) highly similar DNA sequences is called homologous recombination and this kind of recombination plays an important role in recombinational repair. Naturally, eukaryotic chromosomes form synaptonemal complexes during ...., a stage at which the .... crossing-over between the duplexes occurs. This becomes visible in form of the ...., places at which the duplexes are held together during chromosome separation.

Recombination can be initiated by double-strand breaks (both strands in one duplex) or breaks in homologous strands of two duplexes. The latter leads to .... exchange in that the free .... invade the other duplex to pair with the unbroken strands. This creates a crossover point or a 'Holliday junction'. The Holliday junction is resolved by a second nick and if the second nick is in the same strand as the first nick, ... DNA is produced. If the second nick is made in the complementary strand, the situation equals that of double-strand breaks in both duplexes and reciprocal or .... DNA is produced. If recombination is induced by a double-strand break in one DNA duplex, free 3'-OH ends will be generated by .... activity. In bacteria, this may be facilitated by the RecBCD complex. The following single-strand invasion of the healthy duplex displaces one strand comparable to mitochondrial D-loop replication. The displaced strand will serve as .... for synthesis of the other strand in the injured duplex. At this point a double crossover is established and the branchpoint moves toward healthy DNA. Branchpoint movement is powered and carried out by the .... protein complex.

The incorporation of phage DNA into the host bacterial genome also employs reciprocal recombination as described for meiotic crossing-over or double-strand break-induced recombinational repair. However, phage DNA incorporation is an example of .... recombination and it occurs rarely at sites in the bacterial genome different from phage DNA attachment sites (...... sites). In addition, this process usually does not involve any DNA synthesis and the reciprocial strand exchange is not facilitated by the RecA protein. Instead, phage and host factors assemble to form the so-called '.................'.

If the crossing-over between alleles is nonreciprocal or the single strand assimilation is of nonhomologous nature, the segregation of traits will not follow Mendelian rules. This effect is known as ....

Crucial to all processes of DNA modification is the ability of the cell to regulate the tension produced by the helix structure and the degree of coiling of the DNA duplex. This topological state of the DNA is described by the .... and a change in the linking number requires the temporary breakage of at least one strand in the duplex. The enzymes able to break and religate the DNA in order to relax or increase .... are called .... and they are classified by their ability to break and religate one or both strands (Topo I and Topo II, respectively).

Additional reading

Camerini-Otero, R.D. and Hsieh, P. (1995). Homologous recombination proteins in prokaryotes and eukaryotes. *Annu. Rev. Genetics* **29**, 509–552.

Feldman, M.W., Otto, S.P., and Christiansen, F.B. (1996) Population genetic perspectives on the evolution of recombination. *Annu. Rev. Genetics* **30**, 261–296.

Haber, J.E. (1992). Exploring the pathways of homologous recombination. *Curr. Opin. Cell Biol.* **4**, 401–412.

Holliday, R. (1990) History of the heteroduplex. *Bioessays* **12**, 133–142.

Huang, W.M. (1996). Bacterial diversity based on type II topoisomerase genes. *Annu. Rev. Genetics* **30**, 79–108.

Kowalczykowski, S.C. and Eggleston, A.K. (1994). Homologous pairing and DNA strand-exchange proteins. *Annu. Rev. Biochem.* **63**, 991–1043.

Murphy, H.S., Palejwala, V.A., Rahman, M.S., Dunman, P.M., Wang, G., and Humayun, M.Z. (1996). Role of mismatch repair in the *Escherichia coli* UVM response. *J. Bacteriol.* **178**, 6651–6657.

Myers, R.S. and Stahl, F.W. (1994). χ and the RecBCD enzyme of *Escherichia coli*. *Annu. Rev. Genetics* **28**, 49–70.

Rivera, E., Vila, L., and Barbé, J. (1996). The *uvrB* gene of *Pseudomonas aeruginosa* is not DNA damage inducible. *J. Bacteriol.* **178**, 550–5554.

Smith, G.R. (1991). Conjugational recombination in *E. coli* myths and mechanisms. *Cell* **64**, 19–27.

Story, R.M., Weber, I.T., and Steitz, T.A. (1992). The structure of the *E. coli* RecA protein monomer and polymer. *Nature* (London) **355**, 318–325.

Taylor, A.F. (1992). Movement and resolution of Holliday junctions by enzymes from *E. coli*. *Cell* **69**, 1063–1065.

West, S.C. (1996).The RuvABC proteins and Holliday junction processing in *Escherichia coli*. *J. Bacteriol.* **178**, 1237–1241.

Chapter 18: Transposons

Multiple-choice questions

(1) Transposition requires homology between donor and recipient sites.
(a) true   (b) false

(2) IS elements:
(a) are all identical.
(b) encode proteins required for transposition.
(c) are flanked by direct repeats.
(d) cause a duplication of host DNA upon integration.
(e) transpose 10^3 times per element per generation.

(3) The IS elements flanking a composite transposon can be:
(a) in the same orientation.
(b) in the opposite orientation.
(c) both functional.
(d) both nonfunctional.

(4) Replicative transposition:
(a) duplicates the transposon, leaving a copy behind at the old site.
(b) moves the element to a new site, leaving no element at the old site.
(c) requires transposase.
(d) requires resolvase.
(e) includes conservative replication, in which the element moves with conservation of each nucleotide bond.

(5) Nonreplicative transposition:
(a) duplicates the transposon, leaving a copy behind at the old site.
(b) moves the element to a new site, leaving no element at the old site.
(c) requires transposase.
(d) requires resolvase.
(e) includes conservative replication, in which the element moves with conservation of each nucleotide bond.

(6) Precise excision of a transposon:
(a) removes the element and both copies of the duplicated target sequence.
(b) restores the target DNA to its pre-insertion sequence.
(c) occurs more frequently than imprecise excision.

(7) Which is/are the role(s) of transposase in nonreplicative transposition?
(a) to cut out the transposon
(b) to generate a staggered cut at the target site
(c) to move the transposon physically to its new site
(d) to connect the transposon to the staggered nicks at the target site

(8) Transposons of the TnA family generally carry three genes: transposase, resolvase and ampicillin resistance.
(a) true   (b) false

(9) Tn10 elements express high levels of transposase.
(a) true   (b) false

(10) Which statements about the effect of *dam* methylation on Tn10 transposition are true?
(a) Methylation of a site in the inverted repeat of IS10R can block transposase binding.
(b) Methylation of a site in P_{IN} stimulates transcription of transposase.
(c) Tn10 transposition is increased 1000-fold in *dam*⁻ mutants.
(d) Immediately after replication, the methylated sites are hemimethylated, allowing transposase expression and action.

(11) Maize (corn) controlling elements:
(a) are similar in structure and function to bacterial transposons.
(b) can cause a variety of changes in chromosome structure.
(c) can cause a variety of changes in the phenotypes of individual maize (corn) kernels.
(d) are active at distinct times during plant development.

(12) *Ds* elements:
(a) are autonomous transposable elements.
(b) are sites of chromosome breakage.
(c) are similar to *Ac* elements.
(d) have suffered internal deletions.
(e) do not have terminal inverted repeats.
(f) transpose by a nonreplicative mechanism.

Concept questions

(1) Describe two ways in which transposons can cause genome rearrangement.
(2) What happens to the target site upon IS element integration? How does this happen?

(3) What is the relationship between a composite transposon and an IS element?

(4) List the steps required for the insertion of a transposon into a new site.

(5) What is the effect of recombination on (a) DNA located between two direct repeats; (b) DNA located between two inverted repeats?

(6) During what process is a cointegrate formed? What is its structure?

(7) Tn10 elements generally only transpose when their own transposase gene is active (as opposed to using transposase expressed from other Tn10 elements present in the genome). What might be the basis of this *cis* preference?

(8) What is hybrid dysgenesis? What causes it?

Problems

(1) The *tnpR* gene of TnA transposons encodes resolvase, which is required for TnA transposition. Unexpectedly, some mutations in *tnpR* increase transposition frequency rather than decreasing it. Propose an explanation and suggest an experiment to test your hypothesis.

(2) You have a series of Tn10 mutants as listed below. What is the effect of each mutation?

(a) P_{OUT} is mutated so that its activity is 5-fold lower.

(b) P_{OUT} is replaced by a copy of P_{IN}.

(c) *OUT* RNA is mutated to decrease its complementarity to *IN* RNA.

(d) *IN* RNA is mutated to compensate for the *OUT* RNA mutation in mutant (c).

Summary worksheet: fill in the blanks

Transposable elements are DNA sequences that can .... to new locations in the genome. This can affect the genome in a variety of ways: they can cause gene ...., they can .... genes by inserting into them, and their promoters can affect the expression of neighboring genes.

The simplest transposable elements are called .... elements. These contain short .... terminal repeats surrounding a gene encoding ...., the enzyme responsible for transposition. Upon integration into a new site, transposable elements always create a .... repeat of the target site.

.... transposable elements have two .... elements flanking a gene encoding .... resistance. Some elements move by .... transposition, in which a copy of the element is left behind at the original site. This mechanism produces a .... intermediate that contains two copies of the element. The enzyme .... then acts to cause homologous recombination between the two copies of the element. Other elements use .... transposition, in which the element moves to a new site but is not copied. This type of transposition requires .... and .... of the .... site (the .... site remains broken).

Transposition of Tn10 (**is / is not**) tightly controlled by keeping the level of transposase very .... Transposase is synthesized from the .... RNA made by the .... promoter of .... The .... promoter makes the .... RNA, which is .... to part of the *IN* RNA and which is (**more / less**) abundant than the *IN* RNA. .... RNA controls the expression of transposase by hybridizing to the *IN* RNA and preventing its .... Tn10 transposition is also regulated by DNA .... First, ... of a site in P_{IN} prevents .... Also, ... of a site in the right-hand inverted repeat prevents .... binding.

Transposable elements were first identified by the pioneering work of Barbara McClintock, who studied variegation of maize (corn) kernel phenotypes. She defined maize transposons called .... elements. When these elements move they can cause chromosome ...., resulting in acentric fragments. These fragments are lost the next time the cell goes through mitosis, leading to a change in .... elements such as .... are competent to transpose while .... elements like .... are not. These elements have suffered internal ...., and can only transpose when an .... element is present in the genome. Transposable elements have also been identified in *Drosophila melanogaster* on the basis of .... This phenomenon involves a variety of chromosome .... and is observed among the offspring when certain strains of flies are crossed. The aberrations are due to transposable elements known as .... which are activated if the egg does not contain a protein that .... transposition.

Additional reading

Furano, A. and Usdin, K. (1995) DNA 'fossils' and phylogenetic analysis: Using L1 DNA to determine the evolutionary history of mammals. *J. Biol. Chem.* **270**, 25301–25304.

Kleckner, N., Chalmers, R., Kwon, D., Sakai, J., and Bolland, S. (1996). Tn10 and IS10 transposition and chromosomal rearrangements: mechanism and regulation *in vivo* and *in vitro*. In *Transposable elements*, H. Saedler and A. Gierl, eds. (New York, New York: Springer-Verlag), *Curr. Topics Microbiol. Immunol.* **204**, 49–82.

Chapter 19: Retroviruses and retroposons

Multiple-choice questions

(1) **Which of the following are genes found in retroviruses?**
(a) gag   (b) pol   (c) env   (d) onc

(2) **The retroviral LTR:**
(a) is found in the viral genomic RNA.
(b) is generated upon integration into the host chromosome.
(c) contains a strong promoter.

(3) **What happens to the host chromosome at the site of retroviral integration?**
(a) 4–6 nucleotides are deleted.
(b) 4–6 nucleotides are duplicated on one side of the integration and deleted from the other side.
(c) 4–6 nucleotides are duplicated on each side of the integration.
(d) two nucleotides are deleted from the right side.
(e) two nucleotides are deleted from the left side.

(4) **Which of the following are parts of an LTR?**
(a) U3   (b) U4   (c) R   (d) U5

(5) **Retroviral integrase:**
(a) is a site-specific endonuclease.
(b) acts as an exonuclease on the LTR.
(c) generates staggered cuts in the target DNA.

(6) **Delta elements:**
(a) contain active promoters.
(b) can be left behind in genomic DNA when Ty transposes.

(7) ***Copia* elements:**
(a) are present in more than 20 000 copies in the *Drosophila* genome.
(b) are grouped together in tandem arrays.
(c) are flanked by direct repeats.
(d) are not transcribed.
(e) have homology to retroviral *env* genes.

(8) **L1 elements:**
(a) are part of the viral superfamily of retrotransposons.
(b) are found in 20,000 to 50,000 copies per mammalian genome.
(c) are approximately 300 nucleotides in length.

(d) contain an open reading frame with homology to reverse transcriptase.

(e) are transcribed.

(f) have LTRs.

(9) Alu elements:

(a) are part of the viral superfamily of retrotransposons.

(b) are found in 20,000 to 50,000 copies per mammalian genome.

(c) are approximately 300 nucleotides in length.

(d) contain an open reading frame with homology to reverse transcriptase.

(e) are transcribed.

(f) have LTRs.

(10) Which statements about Alu elements are true?

(a) All Alu elements are identical.

(b) The Alu element is derived from the 7SL RNA, which functions in translation.

(c) Alu elements are transcribed by RNA polymerase II.

(d) Alu elements are never found within structural genes.

(e) These elements have a region homologous to viral origins of DNA replication.

Concept questions

(1) What is the difference between a retrovirus and a retrotransposon?

(2) How are the protein products of the *gag* and *pol* genes generated? How is the mRNA encoding the Env protein generated?

(3) Why is a protease important for the retroviral life cycle?

(4) List each step in transforming viral plus-strand RNA into double-stranded DNA.

(5) Upon integration into the host genome, four to six nucleotides of host DNA are duplicated. Why? How is this similar to the insertion of transposons into new sites (Chapter 18)? In addition, two nucleotides are lost from the 5' side of the 5' U3 and from the 3' side of the 3' U5. Does this mean that genetic information is lost from the retrovirus?

(6) Retroviruses can acquire cellular genes such as *onc* genes. How does this happen? Are any retroviral genes affected by the acquisition of cellular genes? How can a retrovirus missing essential genes such as *pol* and *env* replicate?

(7) Describe the structure of the yeast Ty element. How is it related to a retrovirus? Why does it not form infectious particles?

(8)   How would you prove that Ty elements go through an RNA intermediate when they transpose?

(9)   How are FB elements different from *copia* elements?

(10) List four differences between retrotransposons of the viral and nonviral superfamilies.

(11) What evidence suggests that Alu elements could function as origins of DNA replication? What evidence is not consistent with this hypothesis?

Problems

(1)   Students who master the concepts in *Genes VI* often receive job offers from large pharmaceutical companies, who need their expertise in molecular biology. Megalon wants you to join their anti-viral team which is designing new drugs to combat HIV infection. To get the job, you are asked to devise a strategy to prevent HIV gene expression without affecting host gene expression. Using your knowledge of retroviruses, you decide to design drugs that will inhibit certain processes. Describe exactly what drugs you will make, how they will prevent HIV gene expression, and why they will not affect host gene expression.

(2)   T. Jacks, H. Madhani, F. Masiarz, and H. Varmus (*Cell* 55, 447–458, 1988) examined the mechanism of translational frameshifting in rous sarcoma virus (RSV). Shown here is the nucleotide sequence of the *gag–pol* mRNA in the region of the frameshift. The amino acid sequence of the Gag protein is shown in italics above the mRNA sequence and that of the Pol protein is shown below. *** indicates a stop codon.

```
Gag:     Thr    Asn    Leu      ***
    5'- A C A A A U U U A U A G G G A G G -3'
                         Ile    Gly    Arg    :Pol
```

The amino acid sequence of the Gag–Pol fusion protein in the region is Thr Asn Leu Ile Gly Arg.

(a)   Propose a model for translation frameshifting in RSV.

(b)   Jacks *et al.* made mutations in the sequence of this region and determined their effects on frameshifting by measuring the ratio of Gag–Pol fusion protein to Gag protein. Any mutation that decreases frameshifting will decrease this ratio. While most mutations in this region did not strongly affect frameshifting, mutation of any of the three adjacent U nucleotides decreased frameshifting more than 10-fold. How does this observation fit with your model?

(c)   These authors also found that an RNA stem–loop structure in the Pol coding region, located immediately 3' to the site of frameshifting, is required for frameshifting. How do you think this RNA structure contributes to frameshifting?

Summary worksheet: fill in the blanks

The retroviral genome is made of .... When a retrovirus infects a cell, the genome is first copied into .... by the enzyme .... Like all DNA polymerases, this enzyme needs a .... To begin this process, a host .... is used for this role. The .... copy then .... into the host genome to form a ....

Retroviruses contain three genes: ...., ... and .... Viral structural proteins are encoded by the .... and .... genes while the .... gene encodes enzymes needed for viral replication including ...., ... and .... The .... and .... proteins are .... from the same mRNA. In some cases (where the two genes are in the same reading frame), translation of .... requires .... of the .... stop codon. If the two genes are in different reading frames, a ribosome .... is required to synthesize the .... protein. Both these processes are (**efficient / inefficient**). Thus, while .... is made by the normal translation mechanism, ... can only be synthesized as a fusion protein. All viral proteins are .... by a viral .... to generate the mature viral proteins.

Retroviruses can acquire host genes such as .... genes. In the process, these .... viruses lose essential viral genes and become replication-.................... . A .... virus can provide normal viral functions, allowing these viruses to replicate.

Yeasts contain a transposable element called the .... element, which is related to retroviruses. These elements move to new sites in the genome by passing through an .... intermediate. The element contains open reading frames homologous to the retroviral .... and .... genes, although no .... homolog is found. The .... element of *Drosophila* is quite similar.

Retrotransposons of the .... superfamily are similar to retroviruses in several ways. They are flanked by .... and encode proteins with homology to .... or .... However, they do not form .... particles. In contrast, members of the .... superfamily of retrotransposons **(do / do not)** encode any proteins and **(are / are not)** flanked by repeats. Both families are actively transcribed, but the viral elements are transcribed by RNA polymerase II and the nonviral elements by RNA polymerase III. Both families of elements transpose via ...., and both are very .... in mammalian genomes.

Additional references

Brookfield, J. (1995). Molecular evolution: retrotransposon revival. *Curr. Biol.* **5**, 255–256.

Coffin, J. (1996). Retroviridae: the viruses and their replication. In Virology, third edition, B. Fields, ed. (Philadelphia, PA, Lippincott-Raven) pp. 1767–1847.

Kingsman, S. and Kingsman, A. (1996). The regulation of human immunodeficiency virus type-1 gene expression. *Eur. J. Biochem.* **240**, 491–507.

Part 5: The eukaryotic genome

Text chapters

Theme

Chapter 2 introduced the large size and complex architecture of eukaryotic cells. Throughout Part 5, you will learn that the eukaryotic genome is similarly large and complex. You will see many ways in which the genomes of eukaryotes are radically different from the prokaryotic genomes you studied in Parts 2 and 3. First, eukaryotic genomes have much more DNA than expected based on their morphological complexity or on the number of genes thought to be required for life. This 'extra' DNA is found in several forms: simple sequences repeated many times, duplicated genes, and intervening sequences within genes. None of these are prominent features of prokaryotic genomes. Eukaryotes use special mechanisms to organize and condense all this DNA (chromatin) and to ensure that all this DNA is faithfully inherited during cell division (chromosomes).

Chapter 21 describes the C-value paradox: the unexpected finding that eukaryotic genomes contain much more DNA than would be expected. By studying the reassociation kinetics of eukaryotic genomes, you will learn that eukaryotic DNA can be classified into three broad fractions called highly repetitive DNA, moderately repetitive DNA, and unique sequence DNA. Prokaryotic genomes, in contrast, are composed mostly of unique sequence DNA.

Several types of sequence comprise the highly repeated DNA fraction. Satellite DNA is composed of short, simple DNA sequences repeated in tandem thousands of times (Chapter 25). This DNA is not expressed and has no known function. Retrotransposons, described in Chapter 19, make up another component of the highly repeated DNA fraction. Unlike satellite DNA, retrotransposons are transcribed. Recall that mammalian genomes contain up to 50,000 copies of the 6.5 kb L1 element and 300,000 copies of the 300 nucleotide Alu element. Together, L1 and Alu elements take up more than 4×10^8 nucleotides, or greater than 12% of the 3.3×10^9 nucleotide human genome!

The moderately repetitive fraction contains duplicated genes. Some of these are genes whose products are required in large amounts such as rRNA genes. For other duplicated genes, like the globin genes (Chapter 23), individual copies are expressed at different times during development. Unequal crossing-over is an important process for all repeated DNA. These crossovers can amplify the number of copies of repeated genes and maintain homogeneity between copies.

The unique sequence fraction contains the DNA we normally consider 'genes': single-copy open reading frames. However, 'extra' DNA makes up a large portion of this class as well. Here, the extra DNA is in the form of introns (Chapter 22), which are transcribed but subsequently removed by RNA splicing (Chapters 30 and 31). Introns allow the generation of multiple gene products from a single gene through the process of alternative splicing. Introns also have important evolutionary implications. The exon shuffling hypothesis suggests that exons correspond to functional domains of proteins and can be rearranged by DNA recombination between introns. Thus new genes can be assembled from these modular domains instead of having to evolve totally *de novo*.

All this DNA creates special challenges for eukaryotic cells. The DNA must be organized and packaged such that it can be efficiently accessed for transcription, DNA replication and segregation into new daughter cells. Chapter 26 describes how eukaryotic cells solve these challenges. For example, segregation relies upon division of the genome into chromosomes, each containing specialized regions that ensure protection (telomeres) and faithful inheritance (centromeres).

On a smaller scale, DNA is organized into nucleosomes by the highly conserved histone proteins (Chapter 27). Nucleosomes are in turn organized into higher order structures with the help of nonhistone chromosomal proteins. Not only do nucleosomes condense DNA, they also affect its expression in many ways (Chapters 27 and 29).

In addition to the nucleus, two other organelles found in eukaryotic cells carry genetic information (Chapter 24). Both mitochondria and chloroplasts contain circular genomes that encode a prokaryotic-like translation apparatus and a handful of mRNAs. These organelles are thought to be derived from prokaryotic organisms that were captured by larger cells long ago. Now they live in an endosymbiotic relationship with their eukaryotic hosts. Genetic information from both the nucleus and the organelles is required for organelle function. Intriguingly, the nuclear and organelle genomes must co-evolve to ensure that gene products from both genomes can function together.

Finally, our discussion of eukaryotic genomes should note that the nucleotide sequence of the entire *Saccharomyces cerevisiae* genome has recently been determined. This remarkable information provides a complete list of all the genes necessary to be a (small) eukaryote. Computer analysis of the genome sequence finds open reading frames, many of which can be assigned to known genes. Many other open reading frames are yet to be identified. A much larger effort is currently under way to sequence the human genome. This will be a tremendous advance to medical science, but raises ethical questions that have not been addressed before. The knowledge you gain from Part V will help you better understand the challenges and rationale for this monumental project, and to form intelligent opinions about how this information should be used.

Key terms

Use the following terms as a guide for your study and for the completion of the summary at the end of each chapter.

Chapter 20

- ❑ antibiotic resistance gene
- ❑ blunt end
- ❑ clone
- ❑ cloning vector
- ❑ colony hybridization
- ❑ cosmid
- ❑ DNA ligase
- ❑ dot blot

- ❑ expression vector
- ❑ nitrocellulose membrane
- ❑ Northern blot
- ❑ plasmid
- ❑ polymerase chain reaction
- ❑ primer
- ❑ probe

- ❑ reporter gene
- ❑ restriction enzyme
- ❑ reverse transcriptase
- ❑ Southern blot
- ❑ sticky end
- ❑ terminal transferase
- ❑ Western blot

Chapter 21

- ❑ C value
- ❑ chemical complexity
- ❑ $C_o t_{1/2}$
- ❑ fast component
- ❑ highly repeated DNA

- ❑ intermediate component
- ❑ kinetic complexity
- ❑ moderately repeated DNA

- ❑ reassociation kinetics
- ❑ repeated DNA
- ❑ $R_o t_{1/2}$
- ❑ slow component
- ❑ unique sequence DNA

Chapter 22

- ❑ alternative splicing
- ❑ cDNA

- ❑ exon
- ❑ exon shuffling

- ❑ intron

Chapter 23

- ❑ coevolution
- ❑ gene cluster
- ❑ gene conversion
- ❑ gene family
- ❑ gene superfamily

- ❑ nucleolus
- ❑ polymorphism
- ❑ pseudogene
- ❑ replacement site
- ❑ silent site

- ❑ tandem repeats
- ❑ thalassemia
- ❑ unequal crossing over

Chapter 24

- chloroplast genome
- endosymbiont
- maternal inheritance
- mitochondrial genome
- non-Mendelian inheritance
- nuclear petite
- protein import
- *rho⁻* petite
- *rhoᵒ* petite

Chapter 25

- buoyant density
- constitutive heterochromatin
- crossover fixation
- density gradient centrifugation
- euchromatin
- facultative heterochromatin
- hierarchical repeat
- *in situ* hybridization
- minisatellite DNA
- saltatory replication
- satellite DNA
- simple sequence DNA
- unequal crossing-over

Chapter 26

- C-banding
- capsid
- CEN sequence
- centromere
- chromatin
- chromosome
- chromosome puffs
- chromosome scaffold
- euchromatin
- G-bands
- G-rich strand
- heterochromatin
- kinetochore
- nuclear matrix
- nucleoid
- packing ratio
- polytene chromosome
- telomerase
- telomere
- telomeric repeat

Chapter 27

- 10 nm fiber
- 30 nm fiber
- acetylation
- core histones
- core nucleosome
- DNAase I
- euchromatin
- H1
- H2A
- H2A•H2B dimer
- H2B
- H3
- H3₂•H4₂ kernel
- H4
- heterochromatin
- histone modification
- hypersensitive site
- indirect end-labeling
- linker DNA
- methylation
- micrococcal nuclease
- nucleosome
- nucleosome positioning
- phosphorylation
- solenoid

Chapter 20: DNA biotechnology

Multiple-choice questions

(1) **Which statements about restriction enzymes are true?**
   (a) Restriction enzymes are exonucleases rather than endonucleases.
   (b) Restriction enzymes cut DNA at specific sequences called recognition sites.
   (c) A restriction enzyme always cuts DNA to leave the same sequences at the ends.
   (d) Some restriction enzymes cut the two DNA strands at slightly different points within their recognition site to make a 'sticky' end.
   (e) Some restriction enzymes cut the two DNA strands at the same place within their recognition site to make a blunt end.

(2) **Which enzyme requires a primer?**
   (a) restriction enzyme
   (b) terminal transferase
   (c) reverse transcriptase
   (d) DNA ligase

(3) **Which is not a step in the Southern blotting procedure?**
   (a) digestion of the DNA with a restriction enzyme
   (b) ligation of the DNA into a vector
   (c) separation of the DNA fragments on a gel
   (d) transfer of the DNA fragments to a nitrocellulose membrane
   (e) hybridization of the membrane with a labeled probe

(4) **Which kind of cloning vector can hold the largest insert of foreign DNA?**
   (a) plasmid
   (b) cosmid
   (c) yeast artificial chromosome
   (d) phage lambda
   (e) cDNA expression vector

(5) **Reporter genes:**
   (a) replace the coding region of a gene of interest with a coding region that is easily assayed.
   (b) replace the promoter region of a gene of interest with a promoter region that is easily assayed.
   (c) can be used to measure the activity of a promoter.
   (d) can be used to determine when and where a promoter is active.

Concept questions

(1) List all the enzymes, in the proper order, that you would need to clone a cDNA copy of an mRNA into an expression vector.

(2) What is the difference between a clone and a probe?

(3) What is the difference between a cDNA clone and a genomic clone?

(4) What is an advantage of a dot blot over a Southern blot? What is a disadvantage?

(5) At what temperature would you incubate a PCR reaction?

(6) What functional DNA sequences does a YAC require? Why is it advantageous to clone large fragments of DNA?

Problems

(1) Western blotting use antibodies to detect specific proteins within a complex mixture of proteins. Several clever derivations of Western blotting have been devised to detect proteins that bind nucleic acids. Can you guess what a Northwestern blot detects?

(2) You are performing a Southern blot analysis, and have just completed the gel electrophoresis step. The next step in the protocol is to soak the gel in a solution of sodium hydroxide to denature the DNA into single strands. To save time, you decide to skip this step and proceed to transfer the DNA from the gel to the nitrocellulose membrane. You hybridize your membrane with a labeled probe, but your final autoradiogram is blank. What went wrong?

(3) Clone libraries are often prepared by partial digestion of genomic DNA with a restriction enzyme that recognizes a four base sequence (these enzymes cut relatively frequently: once every 256 base pairs on average). The enzyme *Sau*3AI has the recognition sequence 5'-\GATC-3' (only the top strand is shown, the cleavage site is indicated \). This site has the advantage that *Sau*3AI fragments can be cloned into a vector digested with *Bam*HI (5'-G\GATCC-3') or *Bgl*II (5'-A\GATCT-3'). Below is a restriction map showing the *Sau*3AI sites surrounding the *YFG1* gene. List the sizes of restriction fragments that would be generated if *Sau*3AI were allowed to cut this fragment to completion. List the sizes that would result from a partial *Sau*3AI digestion of this fragment. If you want to isolate a restriction fragment containing the intact *YFG1* gene, why is it important to use a library generated by partial digestion?

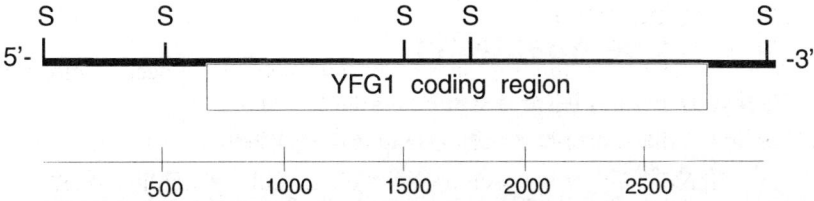

Sau3AI sites are indicated by S; a scale in nucleotides is also shown.

Summary worksheet: fill in the blanks

Cloning relies on the use of many different DNA metabolism ....
that have been isolated (predominantly) from prokaryotic cells.
.... enzymes are required to cut DNA into manageable
fragments. A variety of DNA .... are used including ....
.........., which copies RNA into single-stranded DNA. DNA ....
is used to join restriction fragments together.

Cloning also requires DNA sequences from microbial organisms. These
are called cloning ...., and are used to propagate the cloned
DNA. Plasmid vectors are derived from native bacterial plasmids and
commonly carry genes conferring .... resistance. Other vectors
are made from .... such as λ, which have been modified to carry
foreign DNA. Very large fragments of foreign DNA can be cloned in
vectors called ...., which contain all
the DNA sequences required for chromosome replication and segregation.

To clone a eukaryotic gene of interest, a .... must be made
that will hybridize to the gene (this usually requires some knowledge of
the DNA sequence of the gene). Also needed is a clone .... that
contains many thousands of plasmids (or phages or YACs), each of which
carries a different fragment of eukaryotic DNA. When the probe is
incubated with the clone library, it will .... only to those rare
plasmids that contain a DNA fragment complementary to the probe.

Many powerful and sophisticated techniques have been developed to
analyze cloned DNA. For example, DNA fragments present in very small
amounts can be amplified using the ....
.... Gene activity can be quantified by attaching a ....
gene such as .... or .... to the promoter of the gene of
interest.

Additional reading

Sambrook, J., Fritsch, E., and Maniatis, T. (1989). *Molecular cloning: a laboratory
manual, second edition.* Cold Spring Harbor Press, Cold Spring Harbor, NY.

Chapter 21: Genomes

Multiple-choice questions

(1) **Which statement(s) is (are) true?**
   (a) The C value correlates exactly with the morphological complexity of an organism.
   (b) The C value correlates inversely with the morphological complexity of an organism.
   (c) The minimum C value per phylum correlates roughly with the morphological complexity of an organism.

(2) **$C_o t_{1/2}$ correlates with genome size.**
   (a) true   (b) false

(3) **$C_o t_{1/2}$ correlates with genome complexity.**
   (a) true   (b) false

(4) **The human genome contains exactly enough DNA to encode all the required genes.**
   (a) true   (b) false

(5) **Repetitive DNA consists of:**
   (a) identical repeated sequences.
   (b) one family of related repeated sequences.
   (c) many types of sequence repeated in varying numbers.

(6) **To determine which DNA sequences are transcribed into RNA, a 'tracer' experiment is performed. In this experiment:**
   (a) a large amount of unlabeled RNA is hybridized to a limiting amount of genomic DNA.
   (b) a small amount of labeled RNA is hybridized to excess genomic DNA.
   (c) sufficient excess DNA is used such that all the labeled RNA will hybridize.
   (d) the tracer RNA reassociates with the class of DNA from which it was transcribed.
   (e) most of the tracer hybridizes with moderately repeated sequences.

(7) **When excess RNA is hybridized to limiting DNA:**
   (a) all the RNA hybridizes
   (b) all the DNA hybridizes
   (c) 50% of the RNA hybridizes
   (d) 50% of the DNA hybridizes
   (e) none of the RNA hybridizes, and all the DNA reassociates into double-stranded DNA

(8) **Which of the following are steps in a reassociation experiment?**
   (a) melting the DNA
   (b) reannealing the DNA
   (c) shearing the DNA into fragments
   (d) all of the above
   (e) none of the above

Concept questions

(1) Describe the C-value paradox, including an example of this paradox.

(2) What parameters of a eukaryotic genome affect its $C_ot_{1/2}$ value?

(3) Why does poly(A)–poly(U) reassociate faster than the unique sequence genome of bacteriophage MS2?

(4) Both the T4 and *E. coli* genomes are composed of almost entirely unique sequence DNA, yet the $C_ot_{1/2}$ value of the T4 genome is 37.5 times lower than that of *E. coli*. Why?

(5) How is the stringency of a hybridization experiment controlled?

(6) Association of RNA with genomic DNA has been used extensively to examine the transcription of different classes of DNA. How does a DNA-driven association experiment differ from an RNA-driven experiment?

(7) What does a $R_ot_{1/2}$ experiment measure?

(8) Describe an experiment that measures how much of the unique sequence DNA is transcribed into mRNA (refer to Figure 21.9).

(9) How can the number of transcribed genes be measured? (refer to Figure 21.9).

Problems

(1) You have found a new species of mold growing in an abandoned coffee cup. You know it is a single-celled eukaryotic organism and you want to understand its genome organization.

   (a) What type of experiment would you do to determine the amount and types of repeated DNA in this genome?

   (b) Starting with purified genomic DNA, list each step in the experimental procedure.

   (c) Diagram the expected results, including in your diagram the results from a control prokaryotic genome. Label all components.

   (d) How would you determine what fraction of the unique sequence DNA is transcribed into RNA?

(2) You determine that one component of the genome has a chemical complexity of 2×10^8 and a kinetic complexity of 6×10^5. What kind of DNA is contained in this component?

(3) You have performed a reassociation experiment and separated the fast-reannealing component from the slow-reannealing component. Each component is then individually re-melted. How does the fast-reannealing component behave? How does the slow-reannealing component behave?

(4) A major research effort is currently directed at sequencing the human genome. However, a US Senator recently read Chapter 21 of *Genes VI* and learned that much of the human genome consists of repeated DNA. He thinks it is unwise to sequence the same DNA over and over, and is holding up funding for the project. You receive an emergency call from NIH Director Harold Varmus, who knows you paid careful attention to Chapter 21. The Director wants you to come up with two plans to present before the US Congress. The first plan should ensure that only unique DNA is sequenced. The second plan should allow sequencing of only the unique DNA that is transcribed. Briefly describe your two plans.

Summary worksheet: fill in the blanks

Genome size

The amount of DNA in the genome of an organism is called its .... While lower eukaryotes generally have lower .... than higher eukaryotes, related organisms can differ widely in DNA content. Thus, DNA content does not always correlate with the .... of an organism. An extreme example is found by comparing the genome sizes of flowering plants, where the largest genome contains .... times more DNA than the smallest. It is difficult to imagine that these plants differ so extensively in morphological complexity. In general, higher eukaryotes have much (**more / less**) DNA than expected.

Components of eukaryotic genomes

.... kinetics is used to study the types of DNA found in a eukaryotic genome. The results are generally expressed in a .... curve. In these studies, the genomic DNA is sheared randomly into fragments. The DNA is then heated to .... it into single strands. The reaction is then cooled to allow the single strands to .... and the amount of DNA reassociated at various times is measured. Reassociation depends on two factors: ...., the initial DNA .... and ...., the

110

.... of the reactions. $C_ot_{1/2}$ describes the point in the reaction at which .... % of the DNA has reassociated.

For a simple duplex of homopolymers, such as poly(A)–poly(U), $C_ot_{1/2}$ is (low/high) because any poly(A) strand can pair with any poly(U) strand. For DNA samples with more complex sequences, each strand must find its correct opposite strand, which increases the .... required for reassociation. Therefore, $C_ot_{1/2}$ (increases / decreases) with genome size.

$C_ot_{1/2}$ analysis can also be used to determine genome .... The *E. coli* genome has .... reassociation component which consists of .... sequence DNA. In striking contrast, eukaryotic genomes generally contain .... components which reassociate independently, demonstrating that the organization of the eukaryotic genome is fundamentally different from that of the prokaryotic genome. The .... component comprises roughly .... % of the genome. This component has a (low / high) $C_ot_{1/2}$ value, reassociates (more / less) quickly than the *E. coli* genome, and is composed of .... repeated sequences. The next component to reassociate is called the .... component, representing approximately .... % of the genome. This component has a (low / high) $C_ot_{1/2}$ value, reassociates (more / less) quickly than the *E. coli* genome, and is composed of .... repeated sequences. Finally, the .... component reassociates with a $C_ot_{1/2}$ value more than 10^5 higher than the fast component. Like the *E. coli* genome, this component consists of .... sequence DNA. It reassociates much more slowly than *E. coli* DNA because there are many more different unique single strands that must find their correct opposite strands.

Expression of genome components

Which of these components contain genes that encode mRNA? This can be determined by including radioactively-labeled RNA as a .... in the reassociation experiment. In this 'DNA-driven' experiment, ... is present in vast excess to ensure that all the RNA will hybridize. When hybridization of the labeled RNA is measured, it is found to anneal to the .... and .... components. We therefore conclude that the .... repeated and .... sequence DNA components are transcribed, with most of the transcription coming from the .... component.

Reassociation experiments can measure the expression of nonrepetitive DNA. To do so, the reassociation is performed in .... excess (...........................-driven) and allows calculation of .... values. These data are analyzed by plotting $R_0t_{1/2}$ value versus the proportion of DNA hybridized. Note that the theoretical maximum amount of DNA that can hybridize is .... % of the starting DNA, as only one strand of the DNA is .... to the RNA (the other strand of the DNA is .... to the RNA). Generally, these experiments show that a (small / large) amount of the DNA is able to hybridize to the RNA, indicating that a (small / large) portion of the unique sequence DNA is transcribed.

To study the complexity of the RNA population, RNA can be hybridized to .... In this case, (all / half / none) of the .... will hybridize to the RNA. $R_0t_{1/2}$ is determined not by the number of copies of the genes, but by the number of copies of each RNA present in the RNA sample. As in $C_0t_{1/2}$ analysis, this $R_0t_{1/2}$ analysis detects .... sequence components among the RNA population. In RNA isolated from the chicken oviduct, a large fraction of the cDNA hybridizes rapidly, indicating that the sample contains .... copies of an individual RNA and its cDNA. This is the ovalbumin RNA, which is expressed at very .... levels from a single copy gene. The next component represents seven or eight RNAs that are expressed at .... levels. Finally, the high $C_0t_{1/2}$ component represents .... other mRNAs that are expressed at .... levels.

Reassociation experiments also demonstrate that while all cells have the same set of genes, ... genes are expressed in different cells. While some genes are expressed in all cells (these are called .... genes), others are expressed only in specific cells.

Additional reading

Bernardi, G. (1995). The human genome: origin and evolutionary history. *Annu. Rev. Genetics* **29**, 445–476.

Blattner, F. *et al.* (1997). The complete genome sequence of *Escherichia coli* K-12. *Science* **277**, 1453–1474.

Goffeau, A. *et al.* (1996). Life with 6000 genes. *Science* **274**, 546–567. (This paper reports the complete genome sequence of the yeast *Saccharoymces cerevisiae*.)

Schuler, G. *et al.* (1996). A gene map of the human genome. *Science* **274**, 540–546.

Tatusov, R., Koonin, E., and Lipman, D. (1997). A genomic perspective on protein families. *Science* **278**, 631–637.

Chapter 22: Exons and introns

Multiple-choice questions

(1) **Mark all true statements.**
   (a) Exons are found in the same order in the genome and in cDNA.
   (b) Introns can often be translated.
   (c) All cells in the human body have the same set of genes.
   (d) All cells in the human body express the same set of genes.
   (e) All cells in the human body splice mRNAs for each gene in the same way.

(2) **DHFR genes from various mammals have:**
   (a) short exons
   (b) variable exons
   (c) short introns
   (d) variable introns

(3) **Which statements about the yeast and mammalian genomes are true?**
   (a) Most yeast genes have no introns while most mammalian genes have many introns.
   (b) Most genes in the yeast genome are smaller than most genes in mammalian genomes.
   (c) Most yeast proteins are smaller than the corresponding mammalian proteins.
   (d) Most yeast proteins are about the same size as the corresponding mammalian proteins even though the yeast genes are smaller.

(4) **Alternative splicing can give rise to transcripts that differ in:**
   (a) the 5' untranslated region of the mRNA.
   (b) the coding region of the mRNA.
   (c) the 3' untranslated region of the mRNA.
   (d) all of the above.
   (e) none of the above.

(5) **When comparing genes that are conserved between species, the exons are more likely to be related to one another than the introns.**
   (a) true   (b) false

(6) **Introns are usually found in the same positions in related genes.**
   (a) true   (b) false

(7) **According to the 'exon shuffling' hypothesis:**
   (a) Functional domains of proteins are encoded by single exons.
   (b) New genes arise when DNA recombination brings together new combinations of introns.
   (c) New genes arise when DNA recombination brings together new combinations of exons.
   (d) Evolution is accelerated because new functions (proteins) can be formed by assembling different combinations of exons instead of generating the new function from scratch.

Concept questions

(1) The size of yeast mRNAs generally corresponds to the size of the genes. Mammalian mRNAs, however, are significantly smaller than the genes that encode them. Why?

(2) List three ways in which one gene can encode different proteins.

(3) After a gene duplication, the exons are found to diverge less than the introns. However, the mutations rate for all DNA is the same. Propose an explanation.

(4) What information would you gain from a 'zoo blot'?

(5) What evidence suggests that exons correspond to functional domains of the protein? How does this evidence support the 'exon shuffling' hypothesis?

Problems

(1) Given below are the restriction maps for a region of genomic DNA and for a cDNA derived from it. How many introns are present in the genomic DNA? Where are they?

Genomic DNA

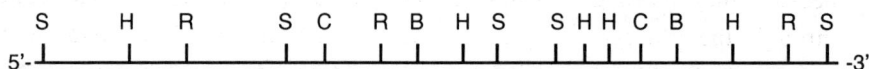

cDNA

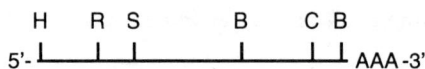

Restriction enzyme abbreviations: B, *Bam*HI; C, *Cla*I; H, *Hind*III, R, *Eco*RI, S, *Sal*I. AAA, poly(A) tail.

114

(2)   You have named your new fungus *Wishida washdit*. You find that *W. washdit* has two actin genes and you determine the DNA sequence of the 5' flanking regions, exon 1, intron 1, exon 2, intron 2, exon 3 and 3' flanking region. In which of these regions do you expect homology between *ACT1* and *ACT2*?

(3)   You have mapped a human disease gene to a 140 kilobase interval of chromosome 7. You have isolated clones for this entire region of the chromosome, but do not know where the gene is located within this region. What information will help you find the gene?

Summary worksheet: fill in the blanks

Prokaryotic genes are colinear with their mRNA products. In striking contrast, eukaryotic genes are much (**smaller / larger**) than their mRNAs. This discrepancy is due to the presence of .... in most eukaryotic genes. Introns are removed by the process of ...., which takes place after .... Although the gene and the mRNA differ in size, the .... of exons within the mRNA does not change.

Introns are (usually) absent from .... genomes, relatively rare in .... eukaryotic genomes, and abundant in .... eukaryotic genomes. Higher eukaryotic exons tend to be .... in length, while introns tend to be .... Intron length also shows greater .... than exon length. The DNA sequence of exons is much (**more / less**) conserved than that of intron sequences.

How did introns arise? One model proposes that introns are ancient and have been lost from .... organisms, for whom rapid DNA replication is an advantage. Introns are maintained in higher eukaryotes because they allow generation of multiple proteins from a single gene through the process of .... Introns have also been proposed to allow the facile generation of new genes through DNA recombination between introns in a process known as ....

Additional reading

Long, M., de Souza, S., and Gilbert, W. (1995). Evolution of the intron–exon structure of eukaryotic genes. *Curr. Opin. Genet. Dev.* **5**, 774–778.

Mattick, J. (1994). Introns: evolution and function. *Curr. Opin. Genet. Dev.* **4**, 823–831.

Chapter 23: Gene numbers

Multiple-choice questions

(1) Which of the following genome features increase as the complexity of an organism increases?
(a) genome size
(b) number of genes
(c) density of genes in the genome
(d) average size of individual genes

(2) Gene duplication:
(a) has occurred many times in eukaryotic genomes.
(b) is critical for generating new genes.
(c) creates gene families.
(d) occurs only within gene clusters.
(e) can occur between clusters located on different chromosomes.

(3) Globin genes:
(a) derive from a common ancestral gene.
(b) are organized into a single cluster.
(c) are organized differently in all higher eukaryotes.
(d) are all expressed simultaneously.

(4) Gene clusters are static and do not change.
(a) true   (b) false

(5) Unequal crossing-over:
(a) takes place within one chromosome.
(b) generates non-reciprocal recombinant chromosomes.
(c) changes gene organization but not total gene number.
(d) begins when chromosomes pair improperly.
(e) reduces gene number at one cluster and increases gene number at another cluster.

(6) Mutations at replacement sites:
(a) do not alter the sequence of the encoded protein.
(b) are found more frequently than mutations at silent sites.
(c) occur at the same rate as mutations at silent sites.

(7) **Which of the following statements about pseudogenes are true?**
  (a) They contain stop codons.
  (b) They are not transcribed.
  (c) They are not translated.
  (d) They can be inactive for any of the above reasons.
  (e) They could eventually be lost from the genome by deletion or other mechanism.
  (f) They could evolve into new genes with different functions.

(8) **Once a gene becomes a pseudogene, replacement sites and silent sites will change at the same rate.**
  (a) true   (b) false

(9) **Pseudogenes are created by unequal crossing-over followed by inactivation of one of the copies. Mark all true statements about this process.**
  (a) The point of inactivation can be determined by comparing the number of changes at silent sites with the number of changes at replacement sites.
  (b) If the pseudogene was inactivated immediately after the gene duplication, it will have more changes at replacement sites than at silent sites.
  (c) If the pseudogene was inactivated a long time after the gene duplication, it will have the same number of changes at replacement sites as at silent sites.

(10) **Which genes are typically organized in tandem units in eukaryotic genomes?**
  (a) globin genes (b) histone genes (c) rRNA genes (d) actin genes

Concept questions

(1) In *Drosophila melanogaster*, what is the connection between the number of genes and the number of bands visible on polytene chromosomes?

(2) What percentage of yeast genes are essential for life? Why might some genes be nonessential?

(3) Which polypeptide chains comprise the hemoglobin molecule in adult mammals? How does alteration of this composition affect the function of the hemoglobin molecule?

(4) How did the two clusters of globin genes arise? How did the globin pseudogenes arise?

(5) Thalassemias are a class of diseases that result from the decreased synthesis of α or β globin. What types of DNA rearrangements can cause these diseases?

(6) Describe how plants, primitive fish, *Xenopus* and mammals provide 'snapshots' of the steps in the evolution of globin gene clusters.

(7) How does a processed pseudogene differ from other pseudogenes? How do processed pseudogenes arise?

(8) Where is the nontranscribed spacer found in the rRNA repeat? Where is the transcribed spacer found? How is the transcribed spacer distinct from an intron?

(9) Why might it be advantageous for the 18S and 28S rRNAs to be transcribed together as a single transcription unit?

Problems

(1) The nucleotide sequence of the entire yeast genome has been determined. Over 70% of the yeast genome is used to encode genes. How will this number differ for the human genome?

(2) In Chapter 22 you found two actin genes in your coffee cup fungus. Now, you continue your study of actin genes in *W. washdit* with several types of experiments.

   (a) First, you probe a Southern blot of *W. washdit* DNA with a labeled probe complementary to a highly conserved region of the actin gene. You find two bands that hybridize strongly with your probe and one band that hybridizes weakly. Propose an explanation for this hybridization pattern.

   (b) You isolate clones for each of these bands. Two correspond to the *ACT1* and *ACT2* genes you have already identified. The third you name *ACT3*. Now you prepare labeled probes specific for each individual actin gene (i.e., they will not cross-hybridize with either of the other actin genes) and use these to probe a Northern blot containing *W. washdit* mRNA. The *ACT1* probe hybridizes strongly with a 2.5 kilobase mRNA, the *ACT2* probe hybridizes weakly with a 2.5 kilobase mRNA, while the *ACT3* probe does not hybridize with any mRNA. Propose an explanation for this hybridization pattern.

   (c) You now determine the nucleotide sequence of *ACT3* and compare all three actin genes. *ACT1* and *ACT2* are identical at 88% of the nucleotides, while *ACT3* is identical to *ACT1* at 43% of the nucleotides. What can you conclude about these actin genes?

118

(3)   Shown below is the sequence of the imaginase gene from
      Saccharoymces cerevisiae (S.c.), *Aspergillus nidulans (A.n.)*, *Drosophila
      melanogaster (D.m.)*, mouse, rat, and human. Identify changes at
      replacement sites and silent sites. Based on the sequence divergence,
      draw a tree showing the history of the imaginase gene.

S.c.	UUA	CCC	AGU	GAU	GAA	CAC	CGU	UUC	AAC	AUG	UAU CUU
A.n.	UUA	CCC	AGU	GAC	GAA	CAU	CGC	UUC	AAC	AUG	UAU CUC
D.m.	UUG	CCC	AGC	GAC	GAA	CAU	CGA	UUU	AAA	AUG	UCC AUC
mouse	CUG	CCU	AGG	GAC	GAA	UAU	AGA	AUU	AAG	AUG	AGU AAG
rat	CUG	CCU	AGG	GAC	GAA	UAU	AGG	AUU	AAG	AUG	AGU AAG
human	CUC	CCU	AGG	GAC	GAA	UAC	AGA	AUC	AAC	AUG	AGC GAC

(4)   You have labeled rRNA and used it as a tracer in a reassociation
      experiment. From which component does it come?

Summary worksheet: fill in the blanks

The human genome contains approximately .... times more
DNA than the *E. coli* genome, yet is estimated to have 125,000 genes, only
.... times more than *E. coli*. Human genes tend to be (**larger /
smaller**) than bacterial or yeast genes, primarily due to the presence of
.... Intriguingly, studies in yeast and *Drosophila* show that (**few
/ many**) genes are essential for the viability of the organism.

      Genes in higher organisms (**are / are not**) all unique; many are present
as members of gene .... When such genes are located adjacent
to one another they form a gene .... Human globin genes are
found in .... clusters called the .... and the
.... Each contains multiple copies of .... globin
genes, which are expressed at different times during ....
Clusters arose by the process of gene .... This occurs when
chromosomes pair .... prior to DNA recombination, in a process
called .... crossing over. This process generates two new
chromosomes known as .... recombinant chromosomes: one has
lost a gene from its cluster while the other has gained a gene. DNA
rearrangements that delete genes from the globin cluster can cause diseases
known as ...., in which insufficient globin proteins are
synthesized.

      During evolution, members of a gene family can change by the
accumulation of .... When these occur at .... sites, the

119

amino acid sequence of the encoded protein is altered. When they occur at .... sites, no change in the encoded protein is caused. For active genes that are under selective pressure, changes will be found more frequently at .... sites. The degree of polymorphism between members of a gene family can be used to estimate the .... since gene duplication.

Some members of gene families, called ...., are not expressed. Frequently, they contain mutations that .... them. Some pseudogenes, called .... pseudogenes, have passed through an .... stage.

Additional reading

Miklos, G. and Rubin, G. (1996). The role of the genome project in determining gene function: insights from model organisms. *Cell* **86**, 521–529.

Henikoff, S., Greene, E., Pietrokovski, S., Bork, P., Attwood, T., and Hood, L. (1997). Gene families: the taxonomy of protein paralogs and chimeras. *Science* **278**, 609–614.

Chapter 24: Organelle genomes

Multiple-choice questions

(1) **What kinds of gene products are encoded in organelles?**
  (a) mRNA
  (b) large subunit rRNA
  (c) small subunit rRNA
  (d) tRNA
  (e) 4.5S rRNA
  (f) 5S rRNA

(2) **How large is a typical chloroplast genome?**
  (a) 1.5 kb   (b) 15 kb   (c) 150 kb   (d) 1500 kb

(3) **How does this compare with the size of the phage T4 genome?**
  (a) 10 times smaller   (b) the same   (c) 10 times larger

(4) **If a plant has 40 chloroplasts per cell, each with 40 copies of the genome, how does the total amount of chloroplast DNA per cell compare with the amount of DNA in a plant nuclear genome? (consider a flowering plant with an average C value; see Figure 21.1)**
  (a) 10 times smaller   (b) the same   (c) 10 times larger

(5) **Organelle genomes:**
  (a) are circular.
  (b) are divided into multiple chromosomes.
  (c) contain large numbers of short repeated DNA sequences.

(6) **Chloroplast genomes contain:**
  (a) two large inverted repeats.
  (b) two large regions of unique sequence DNA.
  (c) two short regions of unique sequence DNA.

(7) **The yeast mitochondrial genome:**
  (a) encodes about the same number of genes as the human mitochondrial genome.
  (b) is about the same size as the human mitochondrial genome.
  (c) contains many introns, some of which can encode proteins.
  (d) contains AT-rich regions.
  (e) has several regions of unknown function.

(8) **In the human mitochondrial genome:**
  (a) almost all the DNA is devoted to encoding gene products.
  (b) almost all protein-encoding genes are transcribed in different directions.
  (c) a single large transcript is made and then cleaved to release individual RNAs.
  (d) most protein-encoding genes are separated by tRNA genes.

(9) **Petite mutants of yeast:**
  (a) have lost all mitochondrial function.
  (b) are always lethal.
  (c) can result from mutation of a nuclear gene that encodes a mitochondrial protein.
  (d) can result from loss or rearrangement of the mitochondrial genome.

Concept questions

(1) Which organelles carry genomes? Why do these organelles carry their own genomes?

(2) How does the mutation rate of mitochondrial DNA differ from that of nuclear DNA? Why should these mutation rates differ?

(3) What aspects of human mitochondrial DNA demonstrate its economical genomic organization?

(4) Can RNA molecules be imported into organelles?

(5) What evidence suggests that organelles are more related to prokaryotes than to eukaryotes?

(6) What happens to the mitochondrial DNA in a *rho⁻* petite strain of yeast?

(7) An extreme case of non-Mendelian inheritance is maternal inheritance, in which a trait is inherited strictly from the mother. Give an example of maternal inheritance in mammals.

Problems

(1) The human nuclear genome contains approximately 3×10^9 nucleotides. The human mitochondrial genome is 16.5 kilobases. For a human cell with 200 mitochondria, what fraction of the total DNA is mitochondrial DNA?

(2) Are human mitochondrial mRNAs polyadenylated? How would you show this?

(3) The human mitochondrial ND6 gene is oriented in the counter-clockwise orientation. The major transcript of human mitochondrial DNA is a full-length clockwise transcript. How does ND6 get transcribed?

(4) What kind of introns would you expect to find in organelle genomes? (You will need information from Chapters 30 and 31 for this problem.)

Summary worksheet: fill in the blanks

Mendel discovered that genes segregate ...., with each parental allele inherited by ....% of the offspring. Segregation of Mendelian traits is governed by the mitotic spindle, which ensures that each daughter nucleus receives one set of chromosomes from each parent. In non-Mendelian inheritance, parental contribution is ....:........ . The genetic information governing non-Mendelian traits is inherited from the .... instead of the ....

Two organelles subject to non-Mendelian inheritance are the .... and the .... Both function in .... conversion, are surrounded by unique .... and carry their own .... Sequencing of organelle genomes reveals the presence of (a few / many) genes, primarily structural .... involved in .... and a few .... Most proteins found in the mitochondria are encoded by .... genes and must be .... into the organelle. The yeast mitochondrial cytochrome *c* oxidase is an enzyme that contains some subunits encoded by .... genes and some encoded by .... genes. It is interesting to speculate how nuclear and mitochondrial genes have evolved together such that their gene products are able to interact to form a functional enzyme.

Studies with inhibitors show that translation inside the organelles is more similar to .... than .... This observation suggests that the mitochondria and chloroplasts in eukaryotic cells are .... They were prokaryotes that were captured and maintained, some of their genes have been transferred to the nucleus.

.... genomes generally encode more gene products than .... genomes, including some subunits of .... polymerase. The chloroplast genome is organized into two .... sequences separated by an extended ....

123

The yeast mitochondrial genome is .... than the human. Many yeast mitochondrial genes contain ....; some of these contain .... In contrast, mammalian mitochondrial genomes are compact and contain no introns. Strikingly, in many cases there is no .... between genes. In fact, some genes even .... by one or more nucleotides, reflecting a remarkable economy of genome use.

All but one human mitochondrial gene are .... in the same .... from a single promoter. The RNAs are derived from a single precursor transcript by .... The genome is organized such that .... genes usually alternate with .... genes. Processing of the .... generates individual mRNAs for the protein-encoding genes.

In yeast, the loss of mitochondrial function is not ...., but does cause a phenotype called .... Without mitochondria, the cells grow (**anaerobically / aerobically**), and their growth rate is lower. Nuclear petites lose mitochondrial function by mutation of a gene encoding an .... protein. Mitochondrial petites of the .... type have lost all mitochondrial DNA while .... petites have suffered a large .... of mitochondrial DNA followed by .... of the remaining sequences.

Additional reading

Grivel, L. (1995). Nucleo-mitochondrial interactions in mitochondrial gene expression. *Crit. Rev. Biochem. Mol. Biol.* **30**, 121–164.

Saccone, C. (1994). The evolution of mitochondrial DNA. *Curr. Opin. Genet. Dev.* **4**, 875–881.

Sugiura, M. (1995). The chloroplast genome. *Essays Biochem.* **30**, 49–57.

Chapter 25: Simple sequence DNA

Multiple-choice questions

(1) Simple sequence DNA

   (a) reassociates with the intermediate component of a $C_o t_{1/2}$ curve.

   (b) consists of individual short repeated sequences that are dispersed throughout the genome.

   (c) accounts for approximately 10% of mammalian genomes.

   (d) has a characteristic buoyant density based on its nucleotide composition.

   (e) is expressed during all phases of the cell cycle.

(2) *In situ* hybridization:

   (a) is a technique in which labeled DNA is hybridized to entire chromosomes and the sites of hybridization visualized by microscopy.

   (b) demonstrates that satellite DNA is dispersed throughout euchromatic regions of the chromosomes.

   (c) reveals that satellite DNA is localized to the centromeres of chromosomes.

(3) More that 40% of the *Drosophila virilis* genome is composed of simple seven base pair sequences that are repeated several million times.

   (a) true   (b) false

(4) Unequal crossing-over:

   (a) occurs rarely between copies of a repeated sequence.

   (b) occurs when repeated sequences mispair.

   (c) results in one repeat cluster growing larger and the other growing smaller.

   (d) takes place in register when an individual hierarchical repeat in one tandem cluster pairs properly with an individual repeat in the other cluster.

   (e) takes place in half-register if one part of a hierarchical repeat from one cluster pairs with a different part of the hierarchical repeat from the other cluster.

(5) Satellite DNA is under strong selective pressure.

   (a) true   (b) false

(6) Minisatellite repeats:

   (a) contain fewer individual repeats per cluster than satellite repeats.

   (b) have a core repeat of 10–15 nucleotides.

   (c) frequently vary in cluster size between individuals in a population.

   (d) are not active in DNA recombination.

Concept questions

(1) Why does satellite DNA form a distinct peak upon density gradient centrifugation?

(2) Why does mammalian satellite DNA produce specific bands when genomic DNA is digested with a restriction enzyme?

(3) What is a hierarchical repeat? How could such a sequence arise?

(4) What is the effect of saltatory replication?

(5) Repeated sequences are not under selective pressure and accumulate mutations rapidly. These properties suggest that repeated sequences should be widely diverged from one another, yet this is not the case. Propose an explanation.

Problems

(1) To examine satellite DNA in *W. washdit*, you perform density gradient centrifugation and prepare a graph showing the amount of DNA in each fraction of the gradient.

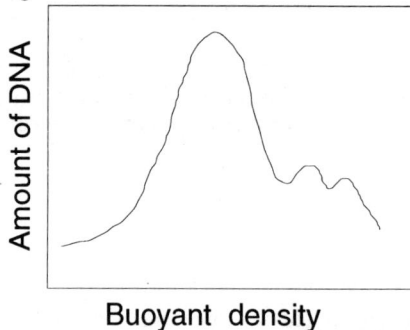

Buoyant density

(a) What DNA components are present in the *W. washdit* genome?

(b) You isolate the major and minor satellite bands, clone a DNA fragment from each, and determine its sequence. Propose an evolutionary history of these repeats.

Major satellite band:

GACATTGGACATGGGACATTGGACATGGAACATTGAACATGGAGCATTGAGCATGG

Minor satellite band:

AGCATGGAGCATGGAGCGTGGAGCGTGGAGCATGGAGCATGGAGCGTGGAGCGTGG

(2) Minisatellite repeats are found throughout the human genome and are highly polymorphic between individuals. Many polymorphic minisatellite repeats have been mapped to near genes associated with human disease. Consider a family in which one parent has diabetes and the other does not. How would you use minisatellite mapping to determine which of their children are at risk for the disease?

126

Summary worksheet: fill in the blanks

Satellite DNA is identified by the technique of ....
.... It differs from the bulk of the DNA due to its
.... Using *in situ* hybridization, satellite DNA is mapped
to regions surrounding the .... This region of the chromosome
is constitutive ...., and (**is / is not**) expressed.

Satellite DNA is predominantly composed of a .... sequence
repeated many times. Mammalian satellite sequences are more complex
than those of lower species and contain a .... of repeating units.
Despite considerable variation, the basic .... base pair repeat can
be fitted into a .... sequence. Satellite sequences have evolved by
successive lateral or .... of the nine base pair repeat
to form larger repeats. Variation is introduced into the repeats by
....

Repeated sequences are prone to .... when chromosomes
pair. This can lead to ...., in which the
number of repeats on one chromosome .... while the number of
repeats on the other chromosome .... The ....
.... theory proposes that identical repeats are created and
maintained by .... until one repeat becomes
....

Minisatellite sequences, like satellite sequences, are composed of
.... repeats. However, minisatellite sequences differ in that they
contain .... repeats, generally .... to ....
copies. The consensus core repeat is .... to ....
nucleotides long and is rich in .... base pairs. The number of
repeats found at a particular minisatellite locus is highly .... in
human populations, making them very useful for ....

Additional reading

Schlötterer, C. and Tautz, D. (1994). Simple sequences. *Curr. Opin. Genet. Dev.* **4**, 823–837.

Chapter 26: Chromosomes

Multiple-choice questions

(1) How much is a small human chromosome compacted when it condenses for mitosis?
  (a) 70-fold
  (b) 700-fold
  (c) 7000-fold
  (d) 70,000-fold

(2) Which strategies do viruses use to get their genetic material inside the capsid?
  (a) They build the capsid around the DNA.
  (b) They build the capsid and then synthesize DNA inside it.
  (c) They build the capsid and then insert the DNA into it.

(3) DNA in bacterial cells is diffusely located throughout the cell.
  (a) true   (b) false

(4) In which cases is the DNA organized into loops?
  (a) viral capsid
  (b) prokaryotic nucleoid
  (c) eukaryotic interphase chromosomes
  (d) eukaryotic metaphase chromosomes

(5) Which features must every eukaryotic chromosome contain?
  (a) telomeres
  (b) centromere
  (c) euchromatin
  (d) an origin of DNA replication

(6) Which statement is always true?
  (a) Condensed genetic material is inactive.
  (b) Decondensed genetic material is active.

(7) Comparison of *CEN* fragments reveals a tripartite consensus sequence. Which parts are essential for *CEN* function?
  (a) CDE-I   (b) CDE-II   (c) CDE-III
  (d) all parts are required for *CEN* function

(8) **Which features of telomerase make it different from other DNA polymerases?**

(a) It synthesizes DNA in a 5' to 3' direction.

(b) The enzyme carries along its own template.

(c) One subunit of the telomerase enzyme is composed of RNA.

Concept questions

(1) What is meant by packing ratio?

(2) List two steps in inserting DNA into a phage head

(3) Describe how the DNA in a prokaryotic cell is organized.

(4) What is the function of the *E. coli* HU protein?

(5) Does DNA associate with the nuclear matrix at specific DNA sequences?

(6) What is the difference between a chromosome scaffold and the nuclear matrix?

(7) What is the difference between a kinetochore and a centromere?

(8) What is the role of topoisomerase II in the scaffold/matrix?

(9) Plasmid inheritance in yeast is random and occasionally, a daughter nucleus does not inherit even a single copy of the plasmid. Plasmid stability is greatly enhanced if the plasmid carries a *CEN* sequence. Why?

(10) What would be the likely consequence of a single point mutation in (a) CBE-II; (b) CBE-III? What proteins bind the *CEN*? What activity is suggested and why is this significant?

(11) What specific challenge faces DNA polymerase in replication of the very end of a chromosome?

Problems

(1) How would you identify a phage lambda mutant defective in DNA packaging?

(2) Ethidium bromide is a small molecule that binds DNA by intercalating between nucleotides. This intercalation introduces positive supercoiling. How does intercalation of ethidium bromide affect the conformation of negatively supercoiled DNA that is (a) free; (b) restrained; (c) bound in domains?

(3) What experimental results would identify specific DNA sequences that are associated with the nuclear matrix? How would you prove that these sequences can bind to the nuclear matrix in the absence of any other DNA sequences?

(4) The telomeric repeat found on the ends of yeast chromosomes is different from that of *Tetrahymena*. What would happen if the gene encoding *Tetrahymena* telomerase RNA were inserted into yeast?

129

Summary worksheet: fill in the blanks

All living cells and viruses must .... their genetic material. Viruses insert their genetic material into a .... Bacterial cells organize their DNA into a dense region known as a .... The DNA is held in large loops that define independent .... The DNA is bound by proteins such as ...., which are thought to condense the DNA into bead-like structures.

Eukaryotic genomes are organized into ...., which condense the DNA and ensure their faithful partitioning to daughter nuclei during .... In interphase, chromosomes are attached to a fibrous network on the inner nuclear membrane called the .... matrix. Mitotic chromosomes consist of a dense network of fibers that anchors loops of DNA. This network is known as a ....

Eukaryotic chromosomes are most readily visualized during ...., when they are .... times more compact than in interphase. .... stays condensed throughout the cell cycle while .... becomes much less condensed. This is thought to facilitate .... heterochromatin is never expressed while .... heterochromatin is selectively inactivated.

Polytene chromosomes of *Drosophila melanogaster* form when a synapsed pair of chromosome undergo .... but do not .... They show a distinctive pattern of .... Some bands transiently expand or ...., due to accumulation of .... Puffs are thought to be sites of active ...., and are therefore thought to represent active genes. Individual genes can be mapped to specific bands by the process of ....

Every chromosome needs three functional regions: a .... for attachment to the mitotic spindle, an .... of DNA replication and .... to prevent loss of genetic information from the chromosome ends. Centromeres contain the ...., to which the microtubules attach during mitosis.

Telomeres have a simple repeated DNA sequence that is rich in .... and .... residues on the 5' end. Different species have (**the same / different**) repeat patterns. ...., the enzyme that synthesizes telomeres, is unusual in that one component is composed of .... This component of telomerase provides the ....

for telomere synthesis. The template is complementary to ....
telomeric repeats. After one full repeat is added, the enzyme ....
forward to add a new repeat. The .... end of the G+T strand
protrudes 14–16 nucleotides past the 5' end of the C+A strand. In this
single-stranded region, G residues form an unusual hairpin structure
known as a ....

Additional reading

Blackburn, E. (1994). Telomeres: no end in sight. *Cell* **77**, 621–623.

Gerasimova, T. and Corces, V. (1996). Boundary and insulator elements in chromosomes. *Curr. Opin. Genet. Dev.* **6**, 185–192.

Pluta, A., Mckay, A., Ainsztein, A., Goldberg, I., and Earnshaw, W. (1995). The centromere: hub of chromosomal activities. *Science* **270**, 1591–1594.

Chapter 27: Nucleosomes

Multiple-choice questions

(1) **How much is the DNA compacted (in length) in the 10 nm fiber?**
(a) 6-fold   (b) 10-fold   (c) 40-fold
(d) 240-fold (e) 1000-fold (f) 10,000-fold

(2) **How much is the DNA compacted in the 30 nm fiber?**
(a) 6-fold   (b) 10-fold   (c) 40-fold
(d) 240-fold   (e) 1000-fold   (f) 10,000-fold

(3) **How much is the DNA compacted in a euchromatic region of a chromosome?**
(a) 6-fold   (b) 10-fold   (c) 40-fold   .
(d) 240-fold   (e) 1000-fold   (f) 10,000-fold

(4) **How much is the DNA compacted in a metaphase chromosome?**
(a) 6-fold   (b) 10-fold   (c) 40-fold
(d) 240-fold   (e) 1000-fold   (f) 10,000-fold

(5) **What is the net charge of histone protein?**
(a) positive   (b) neutral   (c) negative

(6) **What is the overall charge of a nucleosome?**
(a) positive   (b) neutral   (c) negative

(7) **When DNA is coiled around a histone octamer:**
(a) it follows a smooth, circular path.
(b) the DNA enters and exits on opposite sides of the nucleosome.
(c) DNA sequences 40 nucleotides apart are brought into proximity.
(d) all of the above are true.
(e) none of the above are true.

(8) **When new nucleosomes form *in vitro*:**
(a) core histones associate with the DNA one at a time.
(b) a $H3_2 \bullet H4_2$ kernel forms, which binds DNA, followed by the sequential addition of two H2A•H2B dimers
(c) the core octamer forms completely and then binds DNA

(9) **All histones undergo post-translational modification of specific amino acids. These modifications:**
(a) alter the charge of the histone molecule.
(b) are permanent.

(c) occur (in part) on the N-terminal arms of the histones, which are thought to extend out from the core.

(d) occur at specific times during the cell cycle.

(10) **Nonhistone chromosomal proteins are thought to be responsible for the higher order compaction of the 30 nm fiber.**

(a) true   (b) false

(11) **DNAase I hypersensitive sites:**

(a) are approximately 100-fold more sensitive to digestion than normal chromatin.

(b) are regions where the DNA is free of nucleosomes.

(c) are regions where the DNA is free of proteins.

(d) usually occur only when a gene is being expressed.

(e) can be found at promoters, origins of DNA replication, and centromeres.

(f) can be maintained through DNA replication.

(12) **When a gene is active:**

(a) the promoter is generally free of nucleosomes.

(b) the entire gene is generally free of nucleosomes.

(c) the gene is covered by nucleosomes but the chromatin structure is altered in such a way that the entire gene is more sensitive to nuclease degradation.

(13) **Which of the following statements about heterochromatin are true?**

(a) It is transcriptionally inactive.

(b) It remains condensed during interphase, when active chromatin decondenses.

(c) Constitutive heterochromatin is found in specialized parts of the chromosome that are never expressed, such as centromeres and telomeres.

(d) Facultative heterochromatin can control gene expression by placing genes in an inaccessible chromatin structure.

Concept questions

(1) List the contents of a nucleosome, the 10 nm fiber and the 30 nm fiber.

(2) Diagram a nucleosome.

(3) Histones are among the most highly conserved proteins known: comparison of H3 sequences reveals only one amino acid change between sea urchin H3 and calf H3. What does this remarkable evolutionary conservation suggest?

133

(4) Describe an experiment that shows the amount of DNA that is tightly associated with a histone core particle.

(5) According to the X-ray crystal structure of the histone octamer, where does the $H3_2$–$H4_2$ kernel lie in the octamer? Where are the H2A–H2B dimers located?

(6) How is H1 distinct from the other histones? What happens when H1 is removed from chromatin?

(7) Describe an experiment that demonstrates the segregation of histones during DNA replication. When do nucleosomes reform after DNA replication?

(8) What is the effect of acetylation of histones on lysine residues? What is the effect of methylation on lysine residues? What happens when the OH group of a serine residue is phosphorylated? How might these modifications change a nucleosome?

(9) At what point in the cell cycle is H1 phosphorylated? When are the phosphate groups removed from H1? Do these data suggest a role for H1 phosphorylation?

(10) Do nucleosomes lie at specific DNA sequences? How does this affect gene expression?

(11) What evidence correlates DNAase I hypersensitive sites with active genes?

Problems

(1) You digested chromatin from chicken erythrocyte nuclei for 20 minutes at 25°C with the following concentrations of DNAase I: 0, 0.1, 1.0 and 10.0 mg/ml. You loaded these samples, along with molecular weight markers, on a 6% polyacrylamide denaturing gel. Below is shown a diagram of your gel. Unfortunately, you mixed up the samples and cannot be sure what was loaded in which lane of the gel. You are sure that the molecular weight markers are loaded in the left hand lane, but you have forgotten the sizes of the marker bands.

    (a) Which DNAase I concentration was used in each lane? Write your answers in the spaces provided on the left of the diagram.

    (b) Provide approximate sizes for the bands in the marker lane (designated M). Write your answers in the spaces provided on the left of the diagram.

134

Lane	DNAase concentration	Marker size	M	1	2	3	4

1. _____ mg/ml

_____ nt

2. _____ mg/ml

3. _____ mg/ml _____ nt

4. _____ mg/ml _____ nt

_____ nt

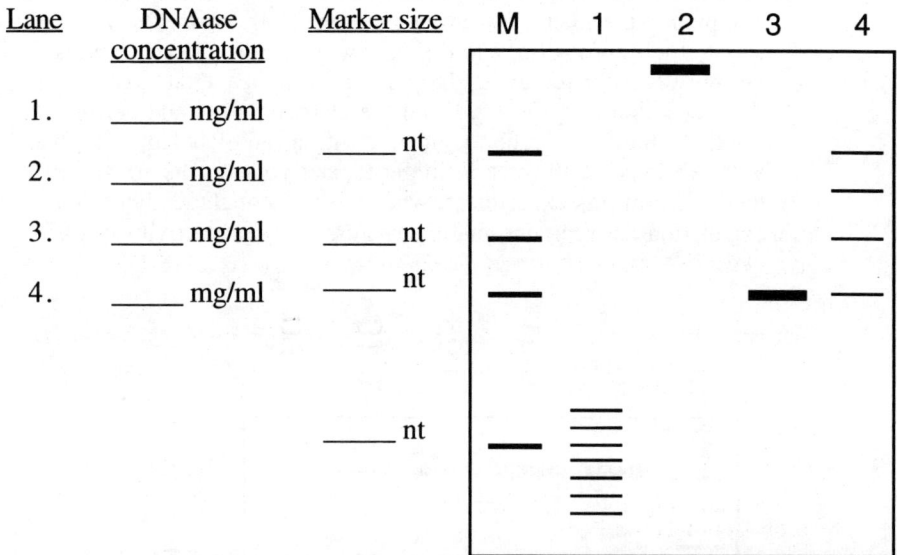

(c) Convince your advisor that your labels are correct by describing the aspects of chromatin structure that are demonstrated by each lane.

(2) Nuclease digestion of chromatin has revealed a great deal about nucleosome structure. In these experiments, the activity of the nuclease is carefully controlled by adjusting both the concentration of the nuclease and the time of digestion. There are many different kinds of these partial digestion experiments that unveil different aspects of chromatin structure. For example, digestion of chromatin with a low concentration of micrococcal nuclease generates a 'ladder' of bands that differ by 200 nucleotides in size. Why is a ladder of bands seen in this experiment? When chromatin is digested with an intermediate concentration of nuclease, a 200 nucleotide band is seen at early times in the digestion. This band is soon trimmed to a 146 nucleotide band. With this amount of nuclease, no ladder is observed. Why not? By using a high concentration of micrococcal nuclease to digest chromatin, a 10 nucleotide ladder is seen. What is the significance of this spacing, and why is a ladder observed here?

(3) To begin your study of chromatin structure in W. washdit, you look at the DNAase I sensitivity of the three actin genes you identified earlier (see Problem 2 in Chapter 23). Recall that by Northern blotting, you have proven that ACT1 produces 90% of the actin mRNA, ACT2 contributes 10% of the actin mRNA, and ACT3 is not expressed.

135

(a) You probe the structure of the actin genes with a DNAase I mapping experiment. You grow a culture of *W. washdit* cells, isolate their chromatin, and digest it with various concentrations of DNAase I. After removing the proteins, you cleave the DNA with the restriction enzyme *Xho*I, run the DNA on a gel and perform a Southern blot. You probe your blot with a radiolabeled probe that will hybridize to all three actin genes, and your results are shown below. From this experiment, what can you conclude about the correlation between chromatin structure and gene activity in *W. washdit*?

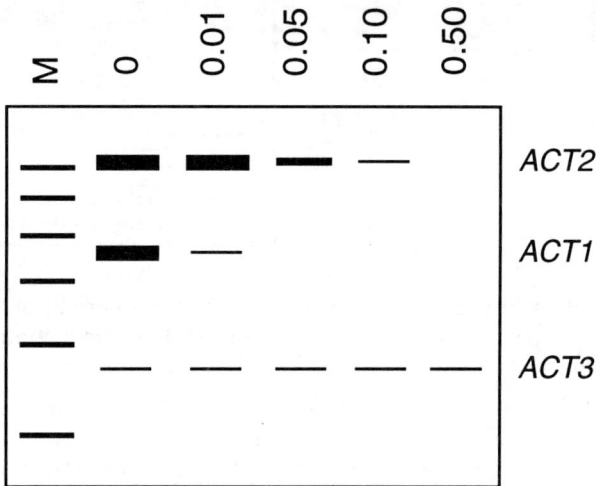

M: molecular weight markers of 1.0 kb, 2.0 kb, 3.0 kb, 4.0 kb, 5.0 kb, and 6.0 kb. The DNAase I concentration (in mg/ml) used is shown above each lane. The *ACT2* gene is found on a 6.2 kb *Xho*I fragment, the *ACT1* gene on a 3.6 kb *Xho*I fragment and the *ACT3* gene on a 1.8 kb *Xho*I fragment.

(b) When you grow *W. washdit* cells, you have always used a rich medium (7.0 M stale coffee pH 2.2, 0.1 M cream, 4.5 M sucrose). A fellow student has told you that cell physiology and gene expression can change when cells are grown under different conditions. You repeat your DNAase I experiment; this time you culture your *W. washdit* cells using a minimal medium that contains only inorganic salts, a nitrogen source and a carbon source. Your results are shown below.

136

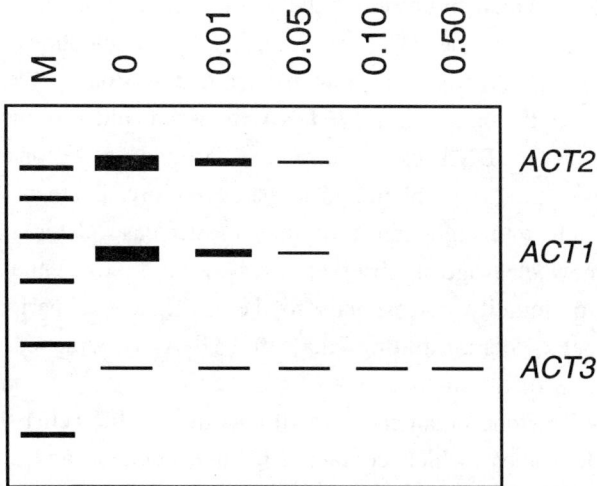

The sizes of the marker fragments and actin fragments are the same as above. The DNAase I concentrations are indicated above each lane.

What change has taken place in the chromatin structure of the actin genes when the cells are cultured in minimal medium? Why does the nuclease sensitivity of *ACT3* not change? Now you want to perform a Northern blot to study actin mRNAs in cells grown in minimal medium. What results do you predict?

Summary worksheet: fill in the blanks

Nucleosome structure

.... are the basic structural unit of chromatin. They consist of a .... core and a length of DNA. The core contains eight ....: two molecules each of ...., ..., ... and .... Histones are .... charged proteins that are highly .... among all eukaryotes. Histones undergo transient post-translational .... including ...., ... and .... All these modifications make the charge of the histones more ...., which is thought to alter their interaction with ....

Nucleosomes in chromatin are spaced about .... nucleotides apart. This was determined by digesting chromatin with a low concentration of .... If a higher concentration of enzyme is used, the chromatin is digested to a length of ....

137

nucleotides, which represents the .... nucleosome. Further digestion reduces the DNA to .... nucleotides. This is the .... DNA, which is protected from digestion by the nucleosome. In contrast, the .... DNA between nucleosomes is readily digested. When DNA is .... around a histone core, DNA sequences .... nucleotides apart are close together.

Digestion with high concentrations of nuclease changes the cleavage pattern: now cleavage is observed every .... nucleotides. This cleavage periodicity corresponds to the .... periodicity of the double helix, demonstrating that the DNA is wrapped around the .... of the histone core.

Isolated histone octamers have **(the same / a different)** shape as from core nucleosomes (which contain a histone octamer and ....). This indicates that nucleosome structure is largely dependent on the .... The structure of the histone octamer has been studied by .... It consists of an .... tetramer and two .... dimers, one .... the tetramer and the other .... it. The path of the DNA in the nucleosome **(is / is not)** smooth.

Gentle extraction of chromatin under low .... (low salt concentration) yields the .... nm fiber. In this form, chromatin resembles .... on a .... At higher ionic strength, the .... nm fiber is seen. This requires the presence of ...., and is a .... of the .... nm fiber, with .... nucleosomes per turn. The 30 nm fiber has a .... ratio of .... Histone H1 **(is / is not)** part of the core particle, and is thought to .... the DNA in the core particle.

Nucleosome formation

Upon DNA replication, nucleosomes reform .... The 'old' histones, which were on the DNA before replication, ... and .... into 'new' nucleosomes. This was determined by growing cells in medium containing .... isotopes of carbon and nitrogen, then switching to medium with .... isotopes prior to DNA .... Chromatin was isolated from the cells and the histone octamers .... to preserve their associations. The .... of the crosslinked histone octamers was determined by .... If the

138

old octamers remained ...., a pool of .... 'old' octamers would be found at the .... of the gradient and a pool of .... 'new' octamers at the .... If the octamers .... and ...., crosslinked octamers of .... density would be found. Nucleosomes can form by two pathways: an .... kernel can form, bind .... and finally bind .... Alternatively, the entire .... can preform prior to DNA binding.

Nucleosome positioning

A critical question is whether nucleosomes are found at specific DNA sequences, and how this could affect gene .... This question of nucleosome .... has been addressed using the technique of .... In this procedure, chromatin is first treated with ...., which cuts .... nucleosomes. The histones are then removed and the DNA digested with a .... DNA fragments are separated by gel electrophoresis and the fragment of interest detected by .... blotting. The fragment of interest will have one end generated by .... and the other end by the .... If the nucleosomes are positioned ...., the .... will always cut the same distance from the .... and the fragment of interest will appear as a .... band. If nucleosomes are positioned randomly, the .... cuts will be various distances from the .... site, and the fragment of interest will appear as a .... band. These experiments always require a control in which .... DNA is treated in parallel. This control detects DNA sequences that the nuclease cleaves ....

Nucleosome positioning can be an .... property of the DNA, as some DNA sequences favor nucleosome formation. Alternatively, nucleosome positioning can be an .... property of the DNA, where nucleosome position is based solely on the distance from other nucleosomes.

Nucleosomes and transcription

Another important question is whether DNA covered by nucleosomes can be .... RNA polymerase must .... the strands of the DNA double helix before it can make RNA; it is not clear how DNA in a nucleosome could be unwound. Experiments with micrococcal nuclease show that actively transcribed genes have **(the same / different)** nucleosome density as/from latent genes. Additional experiments suggest that nucleosomes are .... by RNA polymerase, and .... behind the enzyme after it passes. RNA polymerase moves quickly through the first .... nucleotides of a nucleosome, then begins to .... after the addition of each ribonucleotide. When the enzyme reaches .... through the nucleosome, it speeds up again. This could be the point at which the nucleosome is .... to a new position behind RNA polymerase. Transcription **(can / cannot)** proceed through a crosslinked octamer, indicating that the octamer **(does / does not)** need to dissociate in order to be ....

Very light digestion of chromatin with DNAase I reveals .... These regions are thought to be especially sensitive to digestion because they have **(no / few / many)** nucleosomes. Hypersensitive sites are often found at .... of genes, suggesting that nucleosomes are removed to allow transcription factors access to the DNA. A promoter is hypersensitive to DNAase I only in the .... in which the gene is ....

Although active genes are covered by nucleosomes, they are **(more / less)** sensitive to nuclease degradation than inactive genes. For example, in adult chicken erythrocytes, the adult β-globin gene is degraded at .... concentrations of DNAase I, while the embryonic β-globin gene is digested only at .... concentrations. In some cases, the hypersensitive region extends over more than just the gene. This suggests that chromatin structure is altered in .... containing active genes.

Additional reading

Adams, C. and Workman, J. (1993). Nucleosome displacement in transcription. *Cell* **72**, 305–308

Elgin, S. (1996). Heterochromatin and gene regulation in *Drosophila. Curr. Opin. Genet. Dev.* **6**, 193–202.

Suaren, J. and Hörz, W. (1996). Regulation of gene expression by nucleosomes. *Curr. Opin. Genet. Dev.* **6**, 164–170.

Wolffe, A. and Pruss, D. (1996). Targeting chromatin disruption: transcription regulators that acetylate histones. *Cell* **84**, 817–819.

Part 6: Eukaryotic gene expression

Text chapters

Theme

Part 6 describes the pathway for expression of genetic information in eukaryotic cells. When thinking about eukaryotic gene expression, it is important to keep in mind what you learned about prokaryotic gene expression in Part 3. Also recall that eukaryotic organisms are more complex in cellular structure (Chapter 2) and in organization of genetic information (Part 5). Now, you will see in Chapters 28, 29 and 30, that the eukaryotic gene expression pathway is also more complex. Eukaryotes even rearrange their genetic information to alter gene expression as described in Chapters 32 and 33.

A central point is that eukaryotic RNA polymerases are more complicated and diverse than the prokaryotic RNA polymerase detailed in Chapter 11. Notably, eukaryotes have three separate RNA polymerase enzymes: each one transcribes a different set of genes. Each also contains many subunits, some of which are shared among all polymerases, while others are found only in one of the enzymes. All three polymerases contain subunits that are similar in amino acid sequence to the subunits of prokaryotic RNA polymerase.

None of the eukaryotic RNA polymerases binds directly to DNA. Instead, each polymerase requires that a preinitiation complex of basal transcription factors first assemble on the promoter. For RNA polymerases I and II, these promoters are located entirely upstream of the target gene. Surprisingly, most RNA polymerase III promoters are located within the gene. In all cases, recognition of promoters is more complex in eukaryotes than in prokaryotes, where a single sigma factor directs RNA polymerase to a promoter (Chapter 11).

For RNA polymerase II promoters, additional transcription factors bind to short DNA sequences upstream of the basal promoter to increase promoter efficiency. These 'regulatory' transcription factors increase the efficiency of transcription by forming protein–protein interactions with the basal transcription factors. Some genes also contain enhancer elements,

141

which are located a great distance away from the basal promoter. Enhancers stimulate the activity of the basal promoter by binding additional transcription factors. Thus, eukaryotic transcription relies heavily on the activity of *trans*-acting factors (i.e., proteins) to bring RNA polymerase to a promoter. This also is in contrast to prokaryotic transcription, which is more dependent on *cis*-acting signals within the DNA (Chapters 11 and 12).

A second major point is that eukaryotic genes are almost always interrupted by regions of non-coding DNA called intervening sequences or introns (Chapter 22). Such sequences are not typically found in prokaryotic genomes. This reflects a central feature of eukaryotic genomes; much of the DNA does not encode proteins (Chapter 21). These nuclear pre-mRNAs are spliced by a large, multicomponent complex called the spliceosome. Spliceosomes assemble in a stepwise manner through the association of snRNP particles and free proteins with the pre-mRNA. SnRNPs contain both RNA and protein components; both components are required for snRNP function. The splicing process is very precise and, like transcription, can be regulated to alter gene expression.

Intriguingly, the removal of two other classes of introns, the Group I (Chapter 31) and Group II (Chapter 30) self-splicing introns, is catalyzed by RNA. The splicing of Group I, Group II and nuclear pre-mRNA introns should be considered together because all three reactions proceed through similar two-step mechanisms. In fact, the Group II splicing mechanism is identical to that of nuclear pre-mRNA splicing, raising the possibility that these processes are related through evolution. Furthermore, this shared reaction mechanism suggests that nuclear pre-mRNA splicing is catalyzed by RNA while the protein splicing factors play accessory roles. RNA molecules can catalyze reactions because they are able to adopt complex three dimensional structures (see for example Chapters 5 and 7). These structures can include binding pockets for substrates and cofactors, much as protein enzymes do. However, RNA enzymes, called ribozymes, generally only catalyze reactions involving the phosphodiester backbone of nucleic acids.

In Part 5 you learned that the complex structure of the eukaryotic genome allows gene expression to be controlled in ways not used by prokaryotes, for example, by chromatin condensation. Now you will see in Part 6 that the intricate pathway for gene expression in eukaryotes also offers additional opportunities for the regulation of gene expression. In eukaryotes, gene expression is not simply a question of whether or not a gene is transcribed. Many of the posttranscriptional steps that are unique to eukaryotes (for example 3' end formation, splicing and transport of the mRNA to the cytoplasm) can also be regulated to modulate the structure

142

and amount of the gene product that is produced. Thus, these steps, which are not found in prokaryotes, offer eukaryotes an additional opportunity for regulation.

Finally, eukaryotic cells employ several unusual methods to regulate gene expression that are not used by prokaryotes. The RNA editing process, for instance, allows genetic information to be altered after genes have been transcribed. In addition, eukaryotes can actively rearrange certain DNA sequences to alter gene expression. For example, yeast cells switch mating type by moving mating type genes from silent loci to the active locus. Similarly, trypanosomes rearrange the DNA to create new cell surface proteins that help them evade immune surveillance. The immune system itself uses DNA rearrangement to create a huge number of different antibody genes from a smaller set of gene fragments. Many of these processes use mechanisms similar to the DNA recombination and repair pathways described in Chapters 16 and 17.

In Part 6, you will find several important themes that recur throughout gene expression. One central theme is that gene expression relies on highly specific interactions between biological macromolecules. Transcription factors in mammalian cells, for example, find their proper binding sites within billions of base pairs of DNA. They then form specific contacts with other proteins bound at distant sites. Similarly, during pre-mRNA splicing, the correct splice sites are identified within a pre-mRNA that can be more than one hundred kilobases long. This requires the specific recognition of RNA sequences by proteins and by other RNAs. As you read Part 6, look for the mechanisms by which biological macromolecules achieve this remarkable specificity.

Another important theme is that most stages in gene expression involve the ordered assembly of active complexes. Consider the step wise association of snRNPs and splicing factors with the pre-mRNA, which is reminiscent of the assembly of initiation complexes for RNA polymerase I, II and III. In addition to transcription and nuclear pre-mRNA splicing, processes as diverse as DNA replication (Part 4), 3' end formation (Chapter 30) and translation (Part 2) all take place by the sequential association of multiple factors with their templates.

Key terms

Use the following terms as a guide for your study and for the completion of the summary at the end of each chapter.

Chapter 28

- AAUAAA
- activation domain
- α-amanitin
- assembly factor
- basal promoter
- Box A
- Box B
- Box C
- CF1
- coactivator
- core promoter
- CTD
- DNA-binding domain
- enhancer

- helicase
- initiation factor
- Inr
- kinase
- m7-G
- nucleolus
- Pol II CTD
- preinitiation complex
- RNA polymerase I
- RNA polymerase II
- RNA polymerase III
- SL1
- steroid receptor
- TAF

- TATA box
- TBP
- TFIIA
- TFIIB
- TFIID
- TFIIE
- TFIIF
- TFIIH
- TFIIIA
- TFIIIB
- TFIIIC
- transcription factor
- UBF1
- UCE

Chapter 29

- activation domain
- α-helix
- amphipathic
- bHLH
- CAAT box
- chromatin remodeling
- chromosomal domain
- CpG doublet
- Cys-Cys-Cys-Cys
- Cys-Cys-His-His
- demethylation
- dimer

- DNA-binding domain
- enhancer
- Fos
- GC box
- helix-loop-helix
- hemimethylated DNA
- homeodomain
- HSE
- HSTF
- imprinting
- insulator
- jun

- LCR
- leucine zipper
- MAR
- methylation
- nuclear matrix
- response element
- SP1
- steroid receptor
- SWI–SNF complex
- transcription factor

Chapter 30

- AAUAAA
- alternate splicing
- branch site
- CF I

- CPSF
- Group I intron
- Group II intron
- GT-AG rule

- guide RNA
- hnRNA
- lariat
- poly(A) polymerase

- ❑ ribozyme
- ❑ RNA editing
- ❑ secondary structure
- ❑ snRNA
- ❑ snRNA–pre-mRNA interaction
- ❑ snRNA–snRNA interaction

- ❑ snRNP
- ❑ spliceosome
- ❑ 3' splice site
- ❑ 5' splice site
- ❑ SR protein
- ❑ *trans*-splicing
- ❑ transesterification
- ❑ U2AF

- ❑ U1 snRNP
- ❑ U2 snRNP
- ❑ U4 snRNP
- ❑ U5 snRNP
- ❑ U6 snRNP

Chapter 31

- ❑ apo-lipoprotein B
- ❑ cyclization
- ❑ exogenous G
- ❑ Group I intron
- ❑ Group II intron
- ❑ guide RNA
- ❑ hammerhead

- ❑ intron mobility
- ❑ metal ion
- ❑ P, Q, R, S sequences
- ❑ P1–P9
- ❑ ribozyme
- ❑ RNA editing
- ❑ RNase P

- ❑ secondary structure
- ❑ 3' splice site
- ❑ 5' splice site
- ❑ tertiary structure
- ❑ transesterification
- ❑ viroid
- ❑ virusoid

Chapter 32

- ❑ Agrobacterium
- ❑ antigenic variation
- ❑ a1 protein
- ❑ α1 protein
- ❑ α2 protein
- ❑ a-specific genes
- ❑ α-specific genes
- ❑ basic copy
- ❑ DHFR

- ❑ expression-linked copy
- ❑ gene amplification
- ❑ gene rearrangement
- ❑ HML
- ❑ HMR
- ❑ HO endonuclease
- ❑ MAT
- ❑ mating type

- ❑ PRTF
- ❑ silent cassette
- ❑ SIR gene products
- ❑ STE gene products
- ❑ T-DNA
- ❑ Ti plasmid
- ❑ variable region

Chapter 33

- ❑ allelic exclusion
- ❑ B cell
- ❑ cell-mediated response
- ❑ class switching
- ❑ clonal selection
- ❑ complement protein
- ❑ constant region

- ❑ cytotoxic T-cell
- ❑ D region
- ❑ gene rearrangement
- ❑ heavy chain
- ❑ helper T-cell
- ❑ histocompatibility antigen
- ❑ humoral response

- ❑ J region
- ❑ light chain
- ❑ MHC locus
- ❑ somatic mutation
- ❑ T cell
- ❑ T-cell receptor
- ❑ tolerance
- ❑ transgenic animal

Chapter 28: Initiation of transcription

Multiple-choice questions

(1) Which transcription factors contain TBP?
   (a) TFIIB   (b) TFIIIA   (c) SL1
   (d) TFIID   (e) TFIIIB   (f) UBF1

(2) Which transcription factors are assembly factors?
   (a) SP1   (b) TFIIIB   (c) TFIIH   (d) none of these

(3) Transcription factors have separate domains for DNA binding and transcriptional activation.
   (a) true   (b) false

(4) Enhancers can be located:
   (a) upstream of the promoter
   (b) downstream of the promoter
   (c) near the promoter
   (d) far from the promoter
   (e) in the opposite orientation (i.e. on the other strand) from the promoter

(5) TATA boxes are found at:
   (a) every polymerase II promoter
   (b) most polymerase II promoters
   (c) rare polymerase II promoters
   (d) every polymerase III promoter
   (e) most polymerase III promoters
   (f) rare polymerase III promoters

(6) Phosphorylation of the RNA polymerase II C-terminal domain (CTD) correlates with:
   (a) binding to the preinitiation complex
   (b) the kinase activity of TFIIH
   (c) the presence of specific TAF proteins in TFIID
   (d) the switch from an initiating polymerase to an elongation polymerase
   (e) the release of initiation factors TFIIA, TFIIB, and TFIID

(7) Which of the following statements about TBP are true?
   (a) TBP induces a bend in DNA.
   (b) TBP binds the major groove of the DNA double helix.

    (c) TBP associates with different sets of proteins to recognize diverse promoters.

    (d) TBP is thought to interact with a polymerase subunit that is shared between polymerases I, II and III.

(8) Which of the following conditions can explain why genes are not always active when their transcription factors are present?

    (a) the context of the transcription factor binding site

    (b) the presence of a displacing protein

    (c) the chromatin structure of the transcription factor binding site

    (d) the absence of coactivator proteins

    (e) all of the above

(9) Each transcription factor binding site is recognized by a single transcription factor.

    (a) true   (b) false

Concept questions

(1) Which transcription factors contain TBP? Why are they called positioning factors? Propose a model to explain how all three RNA polymerases can interact with TBP.

(2) What is meant by a preinitiation complex?

(3) RNA polymerase III internal promoters are more than 50 nucleotides downstream of the initiation site. How is RNA polymerase III positioned for correct initiation?

(4) What coding constraints are placed upon RNA polymerase III genes with internal promoters?

(5) When a piece of actively transcribed DNA is damaged, the template strand is repaired first. Propose a model to explain this observation.

(6) Propose an experiment to demonstrate the importance of a TATA box for RNA polymerase II transcription.

(7) List the steps in the initiation of transcription of a 5S rRNA gene. At what point does a stable transcription initiation complex form? How would you determine this experimentally?

(8) How do eukaryotic RNA polymerases I, II and III differ from each other? How are they similar? How do they differ from prokaryotic RNA polymerase? How are they similar?

(9) Histone H2A genes are expressed in all cells while immunoglobulin genes are expressed only in lymphoid cells. Promoters for both genes contain a site for the transcription factor Oct-1, which is present in both cell types. Why are immunoglobulin genes expressed only in lymphoid cells?

Problems

(1) The transcription factor SP1 binds the sequence GGGCGG. Below is shown the sequence of the SV40 early promoter with six of these 'GC' boxes underlined. Also shown are the cleavage sites for the restriction enzymes *Nde*I and *Sma*I. Starting with this DNA fragment and purified SP1, describe in detail how you would use DNAase I to map the SP1 binding sites on the SV40 early promoter. Be sure to describe how the starting material is labeled and include all necessary controls. Draw a picture of a gel showing your results; label all lanes.

```
Nde I                                                                      Sma I
|                                                                          |
CCATATGCCCGCCCCTAACTCCGCCCATCCCGCCCCTAACTCCGCCCAGTTCCGCCCATTCTCCGCCCCGGG
GGTATACGGGCGGGGATTGAGGCGGGTAGGGCGGGGATTGAGGCGGGTCAAGGCGGGTAAGAGGCGGGCCC
```

How can you use this information to discover which sites have the highest affinity for SP1?

(2) Promoters for protein-coding genes in eukaryotic cells contain a basal promoter element that is recognized by RNA polymerase II and a collection of basal transcription factors (e.g., TFIID, TFIIB). However, the basal activity of the promoter by itself is very low and is invariably influenced by other specific transcription factors that bind either to the adjacent upstream promoter region (-120 to -30) or to more distant enhancer sites. Analysis of these regulatory regions indicates that they generally contain many binding sites for different transcription factors that interact with the DNA in a sequence-specific fashion.

When a single factor binding site that makes up part of such a complex regulatory region is analyzed by itself in transfection assays, the activity of the factor binding site is generally low. However, it has been found that linking together multiple copies of the same binding site often results in synergistic promoter activation. The following data indicate the responses for a 20 base pair DNA binding motif for a transcription factor known as FE. This element was cloned upstream of a basal promoter and reporter gene and then introduced into cells in culture for 48 hours:

Number of copies	Reporter gene activity
0	20
1	30
2	135
3	125
4	460

What could account for the pattern of activation observed in this experiment?

In another construct with four copies of the binding site, 30 nucleotides are inserted between each pair of binding sites. This construct is inactive. Why?

(3) While mapping the promoter for *YFG1*, you find that deletions right up to the transcription initiation site do not affect the level of transcription. What does this result suggest about the transcription of *YFG1*? What experiment would you do next?

(4) Clones of the hamster myosin gene that contain 500 nucleotides of upstream sequence are inactive when expressed *in vivo*, but clones containing five kilobases of upstream sequence give full activity. Provide a model to explain this effect.

(5) After RNA polymerase II initiates transcription, the initiation complex needs to shift to an elongation complex. To do so, the polymerase complex must unwind a short region of DNA. On linear DNA, this unwinding requires ATP, TFIIE, TFIIH and helicase activity. However, transcription of supercoiled DNA does not require these factors. Propose an explanation.

Summary worksheet: fill in the blanks

Prokaryotes have .... RNA polymerase while eukaryotic cells have ...., each with a distinct function. Polymerase I makes .... and is localized to a sub-nuclear structure called the .... Polymerase II synthesizes .... and is found in the .... Polymerase III is also found in the .... and synthesizes .... and .... All three polymerases are large protein .... with .... subunits. Several subunits are .... by all three polymerases; others are specific to a single polymerase. The largest subunits are homologous to .... RNA polymerase.

RNA polymerase I

RNA polymerase I promoters are unique in that they are all essentially .... The promoter contains two regions: the .... and the .... Both elements have a similar ...., which is rich in .... pairs. The polymerase I transcription factor .... binds to both regions. Next, the factor .... binds cooperatively. When both factors are bound to the promoter, this is called a ...., which is ready for the addition of RNA polymerase. While UBF1 is a .... polypeptide, SL1 contains .... subunits, including a protein called .... This protein is also required by .... and .... It does not bind directly to the ....; rather, it is thought to function by .... RNA polymerase to the promoter complex.

RNA polymerase III

RNA polymerase III promoters are unusual in that they are usually located .... the genes. Two types of internal polymerase III promoters are found. Type I promoters are generally found in 5S rRNA genes and contain two conserved sequences termed .... and .... Transcription factor .... binds box C and recruits factor .... The next factor to bind is ...., which contains .... and other proteins. This factor binds near the .... site and recruits .... Once .... has bound, the preinitiation complex is .... and does not require the continued presence of .... or .... Therefore, TFIIIA and TFIIIC are termed .... factors while TFIIIB is called an .... factor.

Another type of polymerase III promoter is the upstream promoter, found at some genes encoding .... These promoters may contain the elements PSE, OCT and a .... box. While the .... box is sufficient for transcription initiation, the efficiency of transcription is increased when the other elements are present. In these cases, factor .... may recognize the .... box directly. TFIIIB resembles polymerase I factor .... in three ways. First, it does not bind DNA itself; second, it contains ....; and third, it is thought to interact with RNA polymerase.

RNA polymerase II

RNA polymerase II promoters also consist of several sequence elements, although they are much more diverse than polymerase I or polymerase III promoters. First, a loosely conserved .... (for ....) sequence is found near the transcription start site. Most polymerase II promoters also contain a .... box approximately 25 nucleotides upstream of the start site. This element has

150

the consensus sequence .... and, except for its location, is similar to the prokaryotic .... element. This is the binding site for ...., the first factor to recognize polymerase II promoters. This factor is a large complex composed of .... along with roughly seven other subunits called .... (which stands for ....-................. ....). Different combinations of .... create different forms of ...., which might have different promoter specificities. After TFIID has bound to the TATA box, ... binds upstream of the TATA box, followed by .... This forms the polymerase II .... complex, which now can bind RNA polymerase II and is competent to initiate transcription. However, additional factors .... and .... are required for the release of polymerase II from the promoter. .... contains several enzymatic activities including a protein .... activity. It can phosphorylate the .... (which stands for ....-................. ....) of polymerase II. This phosphorylation is thought to trigger the switch from transcription initiation to transcription .... A minority of polymerase II promoters lack a .... box. .... still binds these promoters, presumably due to association of one of the .... with the Inr.

The Inr and the TATA box constitute the .... polymerase II promoter. In addition to these DNA sequences, many other short DNA sequence elements are usually located .... of the basal promoter and are recognized by specific transcription factors. In contrast to the basal elements, which define the .... of transcription initiation, these upstream elements affect the .... of initiation. Transcription factors bound to upstream elements function by .... with basal factors. Some genes contain additional elements called .... that are located far from the basal promoter and contain a high concentration of .... binding sites. They can be found .... or .... of the promoter, and can even be in the .... orientation. They are thought to interact with the basal promoter when the intervening DNA forms a large .... The net effect is to increase the .... of transcription factors at the promoter.

Transcription factors commonly contain two independent domains: one for DNA .... and one for transcription .... The activation domains can directly contact basal factors including .... or .... Alternatively, they may act indirectly through proteins known as ....

Additional reading

Conaway, R. and Conaway, J. (1993). General initiation factors for RNA polymerase II. *Annu. Rev. Biochem.* **62**, 161–190.

Dahmus, M. (1996). Reversible phosphorylation of the C-terminal domain of RNA polymerase II. *J. Biol. Chem.* **271**, 19009–19012.

Geiduschek, P. and Kassavetis, G. (1995). Comparing transcription initiation by RNA polymerases I and III. *Curr. Opin. Cell Biol.* **7**, 344–351.

Hernandez, N. (1993). TBP, a universal eukaryotic transcription factor? *Genes Dev.*, **7**, 1291–1308.

Ranish, J. and Hahn, S. (1996). Transcription: basal factors and activators. *Curr. Opin. Genet. Dev.* **6**, 151–159.

Sachs, A. and Wahle, E. (1993). Poly(A) tail metabolism and function in eukaryotes. *J. Biol. Chem.* **268**, 22955–22958.

Seroz, T. Hwang, J., Moncollin, V. and Egly, J. (1995). TFIIH: a link between transcription, DNA repair, and cell cycle regulation. *Curr. Opin. Genet. Dev.* **5**, 217–221.

Chapter 29: Regulation of transcription

Multiple-choice questions

(1) **Zinc finger proteins bind zinc:**
   (a) covalently.
   (b) in the presence of DNA.
   (c) by coordination between conserved cysteine and histidine residues.
   (d) in an α-helical region of the protein.

(2) **Zinc finger proteins bind DNA:**
   (a) in the major groove.
   (b) through the C-terminal part of the finger.
   (c) using an α-helical region of the protein.
   (d) by forming two sequence-specific DNA contacts per finger.
   (e) using the conserved amino acids of the finger.

(3) **Steroid receptor transcription factors:**
   (a) all bind the same hormone.
   (b) do not bind DNA in a sequence-specific manner.
   (c) use a consensus sequence to bind zinc different from the one used by the zinc finger.
   (d) dimerize through amino acids in the C-terminal part of the second finger.
   (e) have separate domains for transcription activation, DNA binding and hormone binding.

(4) **Steroid receptors of the glucocorticoid class:**
   (a) are homodimers.
   (b) all bind different palindromic DNA sequences.
   (c) generally all bind the same palindromic DNA sequence with different spacing between the half sites.
   (d) heterodimerize with RXR to bind direct repeats.
   (e) are localized to the nucleus in the presence of their hormone.

(5) **Homeodomain proteins:**
   (a) form a structure with three α helices.
   (b) contact DNA primarily through α helix 3 and an N-terminal arm.
   (c) are similar in structure to prokaryotic helix-turn-helix proteins like the λ repressor.
   (d) are always present in the cell nucleus.
   (e) are important in controlling early development in *Drosophila*.

(6) HLH proteins:

(a) are related in sequence to the helix-turn-helix proteins of prokaryotes.

(b) bind DNA through the loop region.

(c) form two α helices that bind DNA in the major groove.

(d) form amphipathic helices, which place hydrophobic residues on one face of the helix.

(e) none of the above.

(7) bHLH proteins:

(a) have conserved basic amino acids in the loop.

(b) cannot homodimerize.

(c) are always expressed.

(d) interact with HLH through their basic regions.

(e) bind DNA only as heterodimers with HLH.

(f) none of the above.

(8) Which statements about leucine zipper proteins are true?

(a) They bind DNA through conserved leucine residues.

(b) They are similar to HLH proteins in that they have adjacent DNA binding and dimerization domains.

(c) Jun can homodimerize but Fos cannot.

(d) Fos/Jun binds different DNA sequences from Jun/Jun.

(e) Fos/Jun binds DNA more tightly than Jun/Jun.

(9) Mark ALL correct statements.

(a) Genes must be completely demethylated to be expressed.

(b) Active DNA is undermethylated.

(c) The methylation state of a region of DNA can change during development.

(d) Methylation is maintained through DNA replication by an enzyme that recognizes hemimethylated DNA.

(10) The restriction enzymes *Hpa*II and *Msp*I recognize different DNA sequences.

(a) true   (b) false

Concept questions

(1)　In eukaryotes, gene expression can be controlled at many levels not available in prokaryotes. List three of these. Can you think of any more?

(2)　What distinguishes transcription factors of the steroid receptor class from the zinc finger class?

(3)　What is the organization of the DNA recognized by transcription factors of the leucine zipper family? How does the dimeric structure of leucine zipper transcription factors explain the organization of their recognition sites?

(4)　Homeodomain proteins are very different from zinc finger proteins, yet these two families use a similar structural motif to bind DNA. What is it?

(5)　How are genes under coordinate control all activated at the same time?

(6)　List seven ways in which the activity of transcription factors can be regulated. Give an example of each.

(7)　Members of a family of transcription factors share regions of highly conserved amino acids, yet bind different DNA sequences. How is this possible?

(8)　Many transcription factors are cellular proto-oncogenes (genes whose mutant forms are associated with uncontrolled cell division). Why do you think mutant transcription factors might cause oncogenesis? For a hint, see Chapter 37.

(9)　Can transcription factors bind DNA that is covered by nucleosomes?

(10)　Two models can be considered to explain how genes in chromatin are transcribed. In the preemptive model, how do transcription factors and RNA polymerase gain access to the promoter? Why does the dynamic model require ATP?

(11)　Why do mutations in the yeast *SWI* or *SNF* genes affect the transcription of many different target genes?

(12)　Chromosomes are thought to be organized into domains that govern gene expression. What types of functional sites can be found within a domain? How do they act?

(13)　In mammals, genes inherited from the mother and the father can be differentially expressed. What is the basis for this phenomenon?

(14)　MyoD is a bHLH protein that is important in the development of muscle cells. How is its activity regulated?

Problems

(1)　You have mapped a gene causing human disease to a small region of chromosome 1. Fortunately, the DNA sequence of this region has been determined by the genome sequencing project. You find in the DNA sequence a single long open reading frame, a portion of which is shown below.

—Leu-Ala-Cys-Ile-Met-Cys-Glu-Phe-Tyr-Leu-Ala-Ala-Arg-Trp-Lys-Cys-Met-His-Leu-Cys-Lys-Arg-Cys-Trp-Tyr—

(a) What kind of protein does this gene encode?

(b) At the same time, you sequenced several mutants. What can you propose about each one?

(i) —Leu-Ala-Cys-Ile-Met-Cys-Glu-Phe-Tyr-Leu-Ala-Ala-Arg-Trp-Lys-Cys-Met-His-Leu-Pro-Lys-Arg-Cys-Trp-Tyr—

(ii) —Leu-Ala-Cys-Ile-Met-Cys-Glu-Phe-Tyr-Leu-Ala-Ala-Arg-Trp-Lys-Cys-Met-His-Leu-Cys-Lys-Ile-Cys-Trp-Tyr—

(c) Predict the domain structure of this protein.

(2) TFIIIA is a transcription factor required for the expression of 5S rRNA genes. This protein contains nine zinc finger domains, and binds to an internal region of 5S rRNA genes and to 5S rRNA itself.

(a) Describe how you would map the DNA binding sites of the TFIIIA protein.

(b) What *specific* mutations would demonstrate that the zinc fingers are required for DNA binding?

(c) You find that a deletion of 19 amino acids from the C-terminus of TFIIIA binds DNA as well as wild type, but fails to activate 5S rRNA transcription. Propose an explanation.

(d) *Xenopus* oocytes synthesize and store large amounts of 5S rRNA. As 5S rRNA accumulates, TFIIIA binds to it. What effect will this have on transcription of 5S genes? What kind of regulatory mechanism is working here?

(3) Studies of the mouse albumin gene show that an element 10 kb upstream of the gene is required for high expression.

(a) What kind of element is this?

(b) This element contains three binding sites for the transcription factor HNF3. HNF3 binds to this element, both as naked DNA and as chromatin. Shown below is a gel mobility shift assay using substrates of naked DNA (lanes 1–4) or the same DNA on to which chromatin was assembled (lanes 5–8). The same amount of labeled DNA is used in each lane. Each lane shows the substrate incubated with a different amount of purified HNF3.

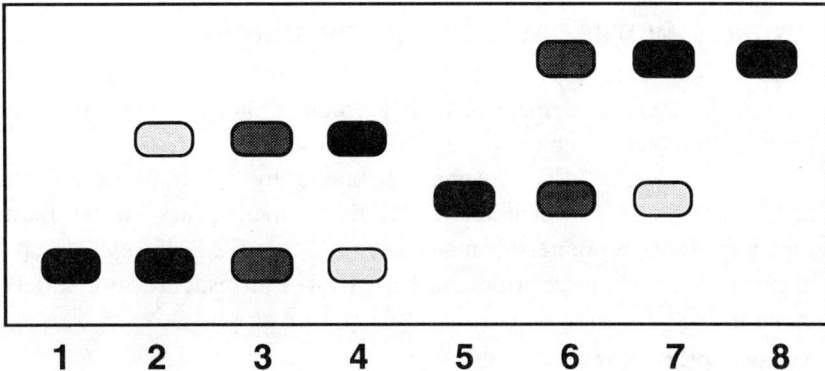

Lane 1, no HNF3; lane 2, 0.1 μg HNF3; lane 3, 0.5 μg HNF3; lane 4, 3.0 μg HNF3; lane 5, no HNF3; lane 6, 0.1 μg HNF3; lane 7, 0.5 μg HNF3; lane 8, 3.0 μg HNF3

What does this result reveal about HNF3?

(c)   How would you determine if HNF3 causes nucleosomes to be specifically positioned?

(d)   Further experiments show that HNF3 binds DNA best when its binding site is positioned exactly in the middle of the nucleosome core. Albumin transcription is increased when the transcription factor GATA4 is present. The DNA element also contains a binding site for GATA4, positioned 42 nucleotides away from an HNF3 site. If the distance between the GATA4 and HNF3 sites is changed to 38 nucleotides or 47 nucleotides, GATA4 does not stimulate transcription. Propose an explanation (a diagram may be helpful).

(4)   Roberts *et al.* studied transcription of the yeast U6 snRNA gene (Roberts, S., Colbert, T. and Hahn S. (1995). *Genes Dev.* **9**, 832–842). This gene has an upstream TATA box, a weak internal A box and a consensus B box that is far downstream. When assayed *in vitro*, both RNA polymerase II and RNA polymerase III can transcribe this gene. *In vivo*, this gene is transcribed by RNA polymerase III and not by RNA polymerase II. How might the polymerase specificity of this promoter be determined?

Summary worksheet: fill in the blanks

Genes that are coordinately regulated often have a conserved .... element in their promoters. For example, heat shock gene promoters contain a DNA sequence known as the .... This element is bound by the transcription factor ...., which allows all heat shock genes to be induced together. Some promoters contain .... different response elements so that transcription can be activated in response to a variety of stimuli.

Transcription factors

The transcription factors that bind these elements are classified into families based on their ....-................ domains. Although each type of DNA-binding domain is distinct, they are responsible for two common features of transcription factors. First, transcription factors usually bind DNA through an .... domain, and second, transcription factors of most families act as ....

Zinc finger proteins bind zinc through a conserved amino acid sequence containing the residues .... Most transcription factors of this family have .... zinc fingers, which contact the DNA through an α-helix formed by the ....-terminal side of the finger. Steroid receptors bind zinc through .... conserved .... residues. DNA binding is carried out by the C-terminal part of the finger while the N-terminal region is involved in .... formation. These factors usually contain three domains: .... activation, ... binding and .... binding. In the case of the glucocorticoid receptor, hormone binding causes a change in the intracellular .... of the protein.

The homeodomain proteins share a 60 amino acid domain that forms three α-helices: the .... one contacts DNA in the .... groove. Transcription factors of this family are important regulators of .... in many organisms.

HLH proteins are not named for a DNA-binding domain. Rather, the helix-loop-helix domain is involved in .... of these transcription factors. This domain forms two .... α-helices separated by a loop region. Many members of this family are called bHLH proteins because they contain a region of .... amino acids next to the HLH domain. This is the DNA-binding domain. bHLH proteins form several types of dimers including homodimers, heterodimers

158

with other bHLH proteins, and heterodimers with HLH proteins. This last class (**does / does not**) bind DNA with high affinity.

The leucine zipper family is similar to the HLH family in several ways. The .... residues form an amphipathic α-helix that is responsible for .... formation. They also bind DNA through an adjacent .... region, and form both .... and .... which have different DNA binding properties.

Effects of chromatin structure

Gene expression is dependent upon the state of ....; most genes covered by nucleosomes (**are / are not**) expressed. Some transcription factors may function by .... histones from promoter region. In contrast, the MMTV promoter must be in a .... for the transcription factors HR, NF1 and OTF-1 to bind. Large-scale chromatin structures called .... can also affect gene expression. Domains are bordered by .... regions and can contain MAR sequences to attach the domain to the .... Within a domain, the .... region contains the promoter and enhancer functions.

Finally, gene expression is also affected when DNA is modified by .... Usually, the .... residue of a .... doublet is methylated. In general, methylated genes are .... Methylation is also responsible for the phenomenon of ...., in which alleles of certain genes from one parent are never expressed.

Additional reading

Buratowski, S. (1995). Mechanisms of gene activation. *Science* **270**, 1773–1774.

Goodrich, J., Cutler, G., and Tijan, R. (1996). Contacts in context: promoter specificity and macromolecular interactions in transcription. *Cell* **84**, 825–830.

Harrison, S. (1991). A structural taxonomy of DNA-binding domains. *Nature* **353**, 715–719.

Nelson, H. (1995). Structure and function of DNA-binding proteins. *Curr. Opin. Genet. Dev.* **5**, 180–189.

Razin, A. and Cedar, A. (1994). DNA methylation and genomic imprinting. *Cell* **77**, 473–476.

Triezenberg, S. (1995). Structure and function of transcriptional activation domains. *Curr. Opin. Genet. Dev.* **5**, 190–196.

Chapter 30: Nuclear splicing

Multiple-choice questions

(1) **Which statements about splice sites are true?**
- (a) Splice sites contain long consensus sequences.
- (b) The 5' splice site and 3' splice site are complementary to each other.
- (c) Almost all splice sites adhere to the GT-AG rule.
- (d) Splice sites are retained in the mature mRNA.
- (e) The 5' and 3' splice sites of an intron can be separated by a large distance.
- (f) The 5' splice site of one intron can work only with the 3' splice site of the same intron: hybrid introns are not spliced.

(2) **The branch site nucleotide:**
- (a) is always an A.
- (b) is located randomly within the intron.
- (c) is found within a loose consensus sequence.
- (d) initiates the first step of splicing by attacking the 3' splice site.
- (e) is covalently joined to three other intron nucleotides after the first step of splicing.

(3) **A transesterification reaction:**
- (a) requires no ATP.
- (b) breaks one bond and forms one bond.
- (c) involves the nucleophilic attack of an OH group on the sugar phosphate backbone.
- (d) all of the above.

(4) **Mark all true statements about snRNAs.**
- (a) snRNAs are strictly localized to the nucleus.
- (b) Most snRNAs are very abundant.
- (c) snRNAs are highly conserved across evolution.
- (d) Some snRNAs can base pair with intron consensus sequences.
- (e) all of the above.

(5) **The snRNPs involved in pre-mRNA splicing:**
- (a) bind a set of proteins known as Sm proteins.
- (b) contain unique proteins (that is, proteins not found on any other snRNP).
- (c) require both their RNA and protein components for function.
- (d) associate with the pre-mRNA primarily through their protein components.
- (e) all of the above.

(6) **Spliceosome assembly:**
- (a) proceeds through an ordered, stepwise pathway.
- (b) involves both snRNPs and soluble proteins (i.e., proteins that are not components of any snRNP).
- (c) needs no ATP.
- (d) is accompanied by multiple rearrangements of snRNPs.
- (e) all of the above.

(7) **SR proteins:**
- (a) are a family of proteins with one or more regions rich in arginine and serine residues.
- (b) all bind different specific RNA sequences.
- (c) form a protein bridge that connects U2AF to the U1 snRNP.
- (d) can bind to splicing enhancer sequences in 3' exons to improve use of weak 3' splice sites.
- (e) can facilitate recognition across exons.
- (f) all of the above.

(8) **Group II splicing and pre-mRNA splicing are similar in that:**
- (a) both types of introns have similar splice sites.
- (b) they both proceed through an identical mechanism involving a lariat intermediate.
- (c) they are both catalyzed by RNA.
- (d) they both require the U1 snRNP.
- (e) they can both take place in the absence of proteins.
- (f) both types of introns form elaborate secondary structures.

(9) **Alternative splicing:**
- (a) uses a completely different mechanism than 'constitutive' splicing.
- (b) allows generation of protein isoforms (variants that differ by addition or deletion of a small number of amino acids) from a single gene.
- (c) involves the use of different 5' and 3' splice sites.
- (d) is used to make different proteins in different tissues and at different developmental stages.
- (e) can involve exon skipping, use of alternative exons, or retention of introns.

(10) **In most cases, splicing takes place only within a single RNA molecule. However, splicing in *trans*:**
- (a) has been observed in both natural and artificial introns.
- (b) involves splicing between a pre-mRNA and a separate SL RNA.
- (c) uses an entirely different splicing mechanism than *cis* splicing.

161

(d) is sometimes used by trypanosomes and nematodes.

(e) puts the same 5' exon (the SL RNA) on to all mRNAs.

(f) requires U1 and U5 but no other snRNPs.

(11) The introns in tRNA:

(a) are removed by a two-step transesterification mechanism.

(b) have a sequence complementary to the anticodon.

(c) all form a similar secondary structure.

(d) are removed by protein enzymes (endonuclease and ligase).

(e) have consensus sequences at the 5' and 3' splice sites.

(12) Addition of the poly(A) tail to a pre-mRNA:

(a) involves a two-step transesterification mechanism.

(b) requires a conserved AAUAAA sequence.

(c) takes place immediately after the AAUAAA sequence has been transcribed.

(d) proceeds by the stepwise assembly of a multicomponent complex.

(e) is carried out by a template-dependent RNA polymerase.

Concept questions

(1) If a branch site is moved to a new location within the intron, it is not active. Instead, a cryptic branch site, near the location of the correct one, becomes activated. Propose an explanation.

(2) List the steps in the spliceosome assembly pathway. Which complex is the active spliceosome?

(3) What is the basis for the interaction between the U1 snRNP and the 5' splice site? How was this demonstrated?

(4) List six RNA–RNA interactions that take place during pre-mRNA splicing.

(5) Genetic screens have isolated many yeast genes whose products are required for pre-mRNA splicing. Several of these genes encode RNA helicase proteins. Why would you expect pre-mRNA splicing to require multiple RNA helicases? When do think these enzymes might function?

(6) What is the catalytic center of the pre-mRNA spliceosome? How is this formed? What evidence supports this model?

(7) Pre-mRNA splicing is thought to have evolved from Group II splicing. What evidence suggests that this is the case? How might this evolution have happened? Why would this be advantageous for the organism?

(8) Describe how alternative splicing controls sex differentiation in *Drosophila*.

162

(9) Compare the process of transcription termination and 3' end formation of products of RNA polymerase I, RNA polymerase III and RNA polymerase II.

(10) What RNA structures are required for processing the 3' ends of histone mRNA? How would you prove that RNA secondary structure, rather than primary sequence, was involved? Can you think of similar structures formed in other reactions?

(11) Why do think it might be advantageous to trypanosomes to put the same first exon (the SL exon) on all its mRNAs?

Problems

(1) Given a multi-intron gene, how would you determine the order in which the introns are removed?

(2) Locate the 3' splice site within this sequence:

5'-ACGUACUAACAUUCUAUUCCUUAAGUUCAUAAGUUGAGUC-3'

If the 3' splice site consensus sequence is mutated, a nearby 'cryptic' 3' splice site is often chosen. Where do you think that would be? What effect will this have on the protein encoded by this pre-mRNA?

(3) Alternative splicing of the *tra* pre-mRNA in *Drosophila* involves a choice between two competing 3' splice sites. The female-specific (FS) 3' splice site is downstream of a stop codon; the non-sex-specific (NSS, used in both males and females) splice site is upstream of this stop codon. Each 3' splice site is preceded by a polypyrimidine region (Py-FS or Py-NSS, see diagram).

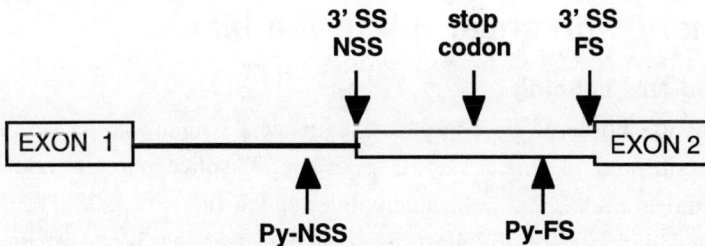

The 3' splice site is chosen when U2AF binds to Py-FS or Py-NSS. Unlike U2AF, which is present in all cells, the Sex lethal protein (Sxl) is present only in females. Sxl is an RNA-binding protein that, like U2AF, binds to pyrimidine-rich RNA. Valcarcel *et al.* (*Nature* **362**, 171–175, 1993) measured the binding affinities of the Sxl and U2AF proteins to Py-FS and Py-NSS. Their results are summarized in the following table:

	U2AF	Sxl
Py-NSS	strong	very strong
Py-FS	moderate	none

Considering these results, propose a specific model that explains how the Sxl protein causes female-specific splicing of *tra* pre-mRNA.

(4)   An extensive series of mutants has been used to study the function of the U6 snRNA. Each U6 mutant listed below prevents splicing. Referring to Figure 30.10, determine at which point each U6 mutant will block spliceosome assembly.

(a)   U6 mutant 1: extends complementarity to U4 by 25 nucleotides;

(b)   U6 mutant 2: reduces complementarity to the 5' splice site to two nucleotides;

(c)   U6 mutant 3: point mutation in the catalytic core;

(d)   U6 mutant 4: extends complementarity to U2 by 25 nucleotides.

(5)   Melissa Moore and Phil Sharp (*Science*, **256**, 992–997, 1992) devised a technique to create a pre-mRNA that contains a single **deoxy**ribonucleotide at a specific position. You have used this technique to make three pre-mRNAs, each of which is composed entirely of ribonucleotides except for a single deoxyribonucleotide at one specific site. The first pre-mRNA has a deoxyribonucleotide at the first nucleotide of the intron, the second has its deoxyribonucleotide at the branch site, and the third has its deoxyribonucleotide at the last nucleotide of the intron. Do you expect these three pre-mRNAs to be spliced?

Summary worksheet: fill in the blanks

Pre-mRNA splicing

Introns are bordered by consensus sequences: .... at the 5' splice site and .... at the 3' splice site. In addition, an essential .... nucleotide, called the .... site, is found within a loose consensus sequence near the 3' end of the intron. Most introns also have a region rich in .... residues located between the branch site and the 3' splice site.

In the first step of splicing, the .... OH group of the branch site adenosine attacks the ...... splice site. This attack breaks the bond between the last nucleotide of the .... and the first nucleotide of the .... Concomitantly, a new ......–5' bond is formed between the first nucleotide of the .... and the

164

.... nucleotide, creating a .... structure. The 5' exon now ends in a 3' .... group, which initiates the second step of splicing by attacking the .... splice site. This attack breaks the bond between the last nucleotide of the .... and the first nucleotide of the 3' ...., while simultaneously forming a new bond between the .... The end products are .... mRNA and released intron, still in the .... form.

The splice sites are recognized by .... particles, which contain both .... and .... components. The snRNPs .... onto the pre-mRNA in a stepwise pathway to form a large multi-component complex called the ...., in which splicing occurs. First, the .... snRNP binds to the .... splice site. Also at this stage, the protein splicing factor .... binds to the .... tract near the 3' end of the intron. Next, the .... snRNP binds to the .... Recognition of the 5' splice site and the branch site relies in part on ....–........... interactions between the intron and the snRNAs. The next step is the addition of the .... trimer to the complex, followed by a series of .... rearrangements. Important changes that occur during these steps include disruption of the .... interaction and formation of a .... pairing. In addition, ... is displaced from the 5' splice site by .... These processes culminate in the creation of an RNA .... center, which can then initiate the first step of splicing.

The identical reaction mechanism is used by Group .... introns, which are autocatalytic. Unlike pre-mRNA introns, Group II introns fold into an elaborate .... structure with .... domains. Domains .... and .... are strikingly similar to the helices formed by .... and ...., respectively, in pre-mRNA splicing. It has been proposed that pre-mRNA splicing .... from Group II splicing, and that spliceosome assembly creates an RNA catalytic center similar to that of Group II introns.

165

Different types of splicing

Some pre-mRNAs use .... splicing to generate mRNAs with different coding capacities. Alternative splicing proceeds through the identical mechanism as 'constitutive' splicing, but different ....
.... are chosen (the splice sites still obey the GT-AG rule). The *Drosophila transformer* gene, for example, uses a constant 5' splice site and one of two alternate 3' splice sites. In males, an upstream 3' splice site is used so that the next exon contains a .... codon, and Tra protein (**is / is not**) produced. In females, a 3' splice site is chosen that is .... of the stop codon, so the Tra protein (**is / is not**) produced.

While most splicing occurs within one RNA molecule, nematodes and trypanosomes can use .... splicing. This splicing occurs when the .... RNA is spliced on to cellular mRNAs. The SL RNA has many features of a .... particle including binding to the same .... proteins bound by the U-series splicing snRNPs. In trypanosomes, the SL RNA carries out the functions of the .... snRNP, which along with U5, is absent from these cells. *Trans*-splicing does use the same ...., ..., and .... snRNPs as *cis* splicing, indicating that the catalytic mechanism is the same in *cis*- and *trans*-splicing.

In contrast, tRNA introns are removed by a completely different mechanism using protein .... and .... enzymes. tRNA introns do not have .... sequences at their splice sites. Rather, the intron is recognized by the .... structure of the pre-tRNA.

3' end formation

RNA polymerase II transcripts are posttranscriptionally .... on their 3' ends. This requires the conserved sequence .... in the 3' untranslated region of the transcript. This sequence is bound by the factor ...., which then signals other .... factors to bind downstream. When a complete cleavage complex is assembled, the transcript is cleaved .... of the AAUAAA. Subsequently, a .... tail is added by the enzyme .... A few RNA polymerase II transcripts, such as the histone H3 mRNAs, are not polyadenylated. The mature 3' end of these RNAs requires the .... RNA, which base pairs with the histone mRNA to direct cleavage.

Additional reading

Adams, M., Rudner, D., Rio, D. (1996). Biochemistry and regulation of pre-mRNA splicing. *Curr. Opin. Cell Biol.* **8**, 331–339.

Madhani, H. and Guthrie, C. (1994). Dynamic RNA–RNA interactions in the spliceosome. *Annu. Rev. Genet.* **28**, 1–26.

Smith, C., Patton, J. and Nadal-Ginard, B. (1989). Alternative splicing in the control of gene expression. *Annu. Rev. Genet.* **23**, 527–577.

Wahle, E. and Keller, W. (1996). The biochemistry of polyadenylation. *Trends Biochem. Sci.* **21**, 247–250.

Wang, J. and Manley, J. (1997). Regulation of pre-mRNA splicing in metazoans. *Curr. Opin. Genet. Dev.* **7**, 205–211.

Chapter 31: Catalytic RNA

Multiple-choice questions

(1) **Group I splicing requires:**
   (a) monovalent cations
   (b) divalent cations
   (c) tetravalent cations
   (d) the U1 snRNP
   (e) an adenosine branch site
   (f) a guanine nucleotide as a cofactor

(2) **Group I introns can use various forms of guanine including:**
   (a) GMP   (b) GDP   (c) GTP
   (d) dGDP   (e) ddGMP (2',3'-dideoxy-GMP)

(3) **During the course of Group I splicing,**
   (a) the exogenous G is covalently added to the 5' end of the intron.
   (b) GTP is hydrolyzed.
   (c) the intron forms a lariat structure.
   (d) the G-binding site is occupied by the exogenous G in the first step and by an intronic G at the 3' splice site in the second step.
   (e) the excised intron can perform additional reactions including two cyclization reactions.

(4) **The Group I intron folds into a complex secondary structure that**
   (a) has nine base-paired regions.
   (b) is very tolerant of mutation.
   (c) forms binding pockets for the exogenous G and for metal ions.
   (d) brings the ends of the intron together.
   (e) undergoes conformational rearrangements during the splicing reaction.
   (f) creates a catalytic core using the P1 and P9 helices.

(5) **The enzyme RNase P:**
   (a) generates the mature 5' end of tRNAs by exonucleolytic activity.
   (b) contains both RNA and protein components.
   (c) requires both components for cleavage *in vivo*.
   (d) requires both components for cleavage *in vitro*.
   (e) adopts complex secondary and tertiary structures to create a catalytic site.

168

(6) **The hammerhead ribozyme:**
    (a) is a small catalytic RNA.
    (b) requires a secondary structure that contains only 17 conserved nucleotides.
    (c) does not require Mg^{2+} ions.
    (d) can be recreated from two separate RNAs, allowing the 'enzyme strand' to cleave the 'substrate strand'.

(7) **In mammalian cells, RNA editing:**
    (a) can change the coding capacity of a gene *after* transcription.
    (b) creates a stop codon in the middle of the apo-B mRNA in intestinal cells but not in liver cells.
    (c) usually makes many changes in each mRNA.
    (d) can change specific nucleotides by deamination.

(8) **In the mitochondria of trypanosomes, RNA editing:**
    (a) is used to alter extensively the coding capacity of many mRNAs.
    (b) makes only a single change in each mRNA.
    (c) adds and deletes G residues.
    (d) uses guide RNAs that base pair with the pre-edited mRNA next to a region that will be edited.
    (e) has guide RNAs with short regions of complementarity to the edited mRNA.
    (f) proceeds in the same direction as transcription.

Concept questions

(1) Compare the second step of Group I splicing with that of Group II splicing.

(2) List four naturally occurring catalytic RNAs.

(3) Can the Group I intron be altered to create other enzymatic activities? Which activities can be created? What does this suggest about the catalytic site of the Group I intron?

(4) Some self-splicing introns contain open reading frames. What kinds of proteins do these encode? How are they related to intron mobility?

(5) Metal ions are important for all catalytic RNAs. Give one example of how a metal ion functions in a catalytic RNA.

(6) What is the role of guide RNAs in RNA editing? What regions of a guide RNA are complementary (a) to the pre-edited mRNA; (b) to the edited pre-mRNA?

169

Problems

(1) You are continuing your investigations into gene expression in the new species of mold that you found growing in your coffee cup. You would now like to determine if the genome of this species contains any Group I introns. Based on your knowledge of the Group I reaction mechanism, you devise a brilliantly simple experiment to detect Group I introns. What is it?

(2) In addition to the P1–P9 base-paired regions, Group I introns can form one additional base-pairing interaction called P10. The nucleotides of the P1 loop are complementary to nucleotides immediately downstream of the 3' splice site. What function do you think this pairing performs?

(3) Group I introns use RNA structures to form binding pockets for their substrates, for metal ions, and for the exogenous G nucleotide. Michel *et al.* (*Nature*, **342**, 391–395, 1989) discovered the location of the G-binding site within the Group I intron secondary structure. They began their search by making assumptions about the nucleotides that make up the G-binding site. They made one assumption about the conservation of these nucleotides, and another about the effect mutations of these nucleotides would have on G binding and on splicing. What specific assumptions do you think they made regarding these properties of the G-binding site nucleotides?

(4) As indicated in Figure 31.12, the hammerhead ribozyme can be constructed from two separate RNA molecules if they have the proper sequences and pair in the proper way. The top strand, called the 'substrate strand', needs to contain the sequence 5'-GUN-3' (where N is any nucleotide). The bottom strand, or 'enzyme strand', needs to have the sequence of the ribozyme catalytic core, and it needs to base pair with the substrate strand. Then, the enzyme strand will cleave the substrate strand on the 3' side of the N residue. Cellular RNAs that are not normally targets for hammerhead ribozyme cleavage might be degraded by supplying an appropriate enzyme strand. This opens up the possibility of using enzyme strands as therapeutic agents to prevent the expression of unwanted genes (for example, several research groups are designing enzyme strands that can cleave HIV RNAs). How would you design an enzyme strand hammerhead ribozyme to cleave HIV RNA? How would you select a specific target sequence within the HIV RNA? What properties should your enzyme strand have? What properties should it not have? What challenges will you have to meet in order to use RNA as a drug?

(5) Below is shown a portion of the *Trypanosoma brucei coxIII* gene:

5'-GGGAGGTTAAGA-3'

The corresponding region of the fully edited mRNA has the sequence:

5'-GGUUUUGUUUUUUAUUGGUAUUUUUUAGA-3'.

What is the sequence of the guide RNA that edits this region of *coxIII*?

170

Summary worksheet: fill in the blanks

Group I splicing

Group I splicing takes place in .... steps. First, the ....
OH group of an .... residue attacks the
.... side of the first nucleotide of the intron. As a result, the
exogenous .... residue is .... to the intron and the first
.... is liberated. Exon I now has a free .... group on its
.... end. In the second step, this group attacks the ....
side of the .... nucleotide of the downstream exon. This reaction
breaks the bond between the .... and the downstream exon and
forms a new bond between the two ....

All Group I introns adopt a common .... structure with
.... base-paired helices named .... The RNA
sequence of most helices can .... as long as there are
.... changes in the opposite strand. However, in some places, the
.... is conserved, indicating that these nucleotides are
.... for Group I intron function. The conserved sequences
.... and .... base pair to form the .... stem
while sequences .... and .... pair to form the
.... stem. Along with the P3 and P6 helices, P4 and P7 form the
.... core of the group I intron.

Some introns contain .... These
encode proteins that allow the intron to .... to new sites in the
genome. These include an ...., which makes a double-stranded
break in the target DNA, and a ...., which copies the intron into
DNA.

Ribozymes

In addition to splicing, RNA molecules can .... other reactions.
Like proteins, RNAs adopt complex secondary and tertiary structures that
form .... pockets for .... ions, cofactors and substrates.
However, ribozymes generally have a .... catalytic rate than
protein enzymes.

Other catalytic RNAs function as site-specific ....
.... is an enzyme that acts in tRNA processing. Like Group I and
Group II introns, it contains both .... and ....
components *in vivo*. However, *in vitro*, the .... alone is active.

The .... is an RNA motif that cleaves multimeric RNA genomes during replication of .... and .... This forms an RNA structure that positions a .... ion at the catalytic site; this ion participates in catalysis by attacking the cleavage site.

RNA editing

RNA editing is an unusual set of reactions that alter the sequence of a mRNA .... transcription. This reaction **(is / is not)** highly specific. For example, in mammalian intestinal cells, a single .... residue in the apo-lipoprotein B mRNA is changed to a .... residue, creating a .... codon. This occurs by .... of the C residue.

Editing in trypanosome .... is much more extensive, involving .... and .... of multiple .... residues. This editing is directed by .... RNAs, which have three distinct regions. On the 5' end, the guide RNA is complementary to a(n) **(edited / pre-edited)** region of the mRNA. This sequence anneals to the pre-edited mRNA. The central region of the guide RNA is complementary to a(n) **(edited / pre-edited)** region of the mRNA. This region directs the addition and deletion of U residues to create the .... sequence. At the 3' end, guide RNAs have a series of .... residues. New evidence suggests that editing does not proceed by transesterification, but instead relies on protein enzymes to cleave the pre-edited mRNA, add or remove U residues, and ligate the pieces back together.

Additional reading

Cate, J., Gooding, A., Podell, E., Zhou, K., Golden, B., Kundrot, C., Cech, T., and Doudna, J. (1996). Crystal structure of a group I ribozyme domain: principle of RNA packing. *Science* **273**, 1678–1685.

Gold, L. (1990). Catalytic RNA: a Nobel prize for small village science. *New Biol.* **2**, 1–4.

Kable, M., Heidmann, S., Stuart, K. (1997). RNA editing: getting U into RNA. *Trends Biochem Sci.* **22**, 162–166.

Pace, N. (1992). New horizons for RNA catalysis. *Science* **256**, 1402–1403.

Scott, W., and Kulg, A. (1996). Ribozymes: structure and mechanism in catalytic RNA. *Trends Biochem Sci.* **21**, 220–224.

Seiwert, S., Heidmann, S., and Stuart, K. (1996). Direct visualization of uridylate deletion *in vitro* suggests a mechanism for kinetoplastid RNA editing. *Cell* **84**, 831–841.

Chapter 32: Rearrangement of DNA

Multiple-choice questions

(1) **Yeast mating type genes are master regulatory genes that control the expression of many other genes. Mating type genes:**
- (a) can be one of two alleles: **a** or α.
- (b) are found at three separate loci in the yeast genome.
- (c) are expressed from three separate loci in the yeast genome.
- (d) can be silent or expressed depending on the locus at which they reside.
- (e) all of the above.

(2) **The mating type switching reaction:**
- (a) involves moving a DNA cassette from a silent locus (*HML* or *HMR*) to the *MAT* locus.
- (b) usually results in a change in mating type.
- (c) does not occur in haploid cells.

(3) **Which statements about the products of the *MAT* genes are true?**
- (a) The $\alpha 2$ protein interacts with PRTF to repress **a**-specific genes.
- (b) The $\alpha 1$ protein interacts with PRTF to repress α-specific genes.
- (c) The $\alpha 2$ protein interacts with the **a**1 protein to repress haploid-specific genes.
- (d) The $\alpha 2$ protein binds the same DNA sequence (with different spacing) depending on its interaction partner.

(4) **The silent loci:**
- (a) differ from the *MAT* locus by the presence of silencer regions.
- (b) are kept transcriptionally inactive by the products of the *SIR* genes.
- (c) have several DNAase I hypersensitive sites.
- (d) are associated with origins of DNA replication.
- (e) act as recipients in mating type switching.
- (f) are kept transcriptionally inactive by chromatin structure.

(5) **The *HO* gene is required for mating type switching. This gene:**
- (a) encodes a site-specific DNA endonuclease.
- (b) has a promoter that contains many different response elements.
- (c) is expressed at all times in all yeast cells.
- (d) encodes a product that can cleave the silent loci but not *MAT*.

(6) Trypanosomes:
- (a) are unicellular parasites that alternate between insect and mammalian hosts.
- (b) are covered by a coat composed of variable surface glycoprotein (VSG) molecules.
- (c) can express several VSG genes at one time.
- (d) lose their VSG coats while in the mammalian host.
- (e) can escape host immune response by changing their VSG every 1–2 weeks.
- (f) can make ten different VSG molecules.

(7) Which of the following statements about VSG genes are true?
- (a) VSG gene products are immunogically cross-reactive.
- (b) The trypanosome genome contains many VSG genes.
- (c) The VSG genes are all clustered together on one chromosome.
- (d) They can reside near telomeres or in internal regions of a chromosome.
- (e) All telomeric copies are transcriptionally active; all internal copies are transcriptionally inactive.

(8) Which statements about the transcriptional activity of VSG genes are correct?
- (a) Only the VSG gene in the expression site is transcribed.
- (b) The expression linked copy (ELC) is always near a telomere.
- (c) The VSG gene in the expression site can be changed by copying a new internal basic copy gene to the expression site.
- (d) Telomeric VSG genes can be activated without gene duplication.
- (e) In rare cases, a novel VSG gene is created in the expression site by assembling fragments from various basic copy VSG genes.
- (f) all of the above.

(9) The VSG gene in the ELC has an unusual structure including:
- (a) a 5' barren region that contains no restriction enzyme cleavage sites.
- (b) a 3' barren region composed of a simple sequence repeated many times.
- (c) both of the above.
- (d) neither of the above.

(10) Ti plasmids:
- (a) can be transferred from the bacterium *Agrobacterium* into plant cells.
- (b) are transferred as double-stranded DNA.
- (c) cause tumors in plants.

(d) direct the synthesis of opines, which function as food for the bacterium and growth hormones for the plant.

(e) require the bacterial *vir* genes for transfer.

(f) are maintained as extrachromosomal plasmids in the plant cell.

Concept questions

(1) It can be argued that if you understand the yeast mating type system, you understand many of the fundamental principles of molecular biology. Explain how this system exemplifies DNA rearrangement, control of gene expression by chromatin structure, maintenance of chromosomal domains, protein–protein interactions between transcription factors, cell type-specific gene expression, and gene activation by signal transduction.

(2) Describe the domain structure of a VSG protein.

(3) In what ways does surface antigen variation of trypanosomes resemble mating type switching of yeast? How do these processes differ?

(4) Describe the bacterial and plasmid genes needed for T-DNA transfer.

(5) What does methotrexate do to mammalian cells? How do cells become resistant to this drug? What is the difference between stable and unstable resistance?

(6) Describe two techniques by which exogenous DNA can be stably introduced into mammalian genomes.

Problems

(1) The T-DNA undergoes two DNA rearrangement events. First, it is excised from the Ti plasmid. After transfer to the plant cell, it integrates into the plant genome. How are these processes similar? What happens to the left and right repeats?

(2) How do the VirA and VirG proteins regulate expression of other *vir* genes?

(3) Four *SIR* (silent information regulator) genes are required to repress expression of mating type genes at HML and HMR. Haploid cells with mutations in one of these genes cannot mate. Why not?

(4) The yeast *RAD52* gene is required for the repair of double-strand breaks in the chromosomes. Haploid cells with mutant *rad52* die unless they also have a mutation in the *ho* gene. Why do you think that is so?

(5) The mechanism by which trypanosomes change the expressed VSG gene is not understood. One way to study this process would be to isolate a mutant trypanosome that was unable to switch the VSG gene it expresses. How would you isolate such a mutant?

175

Summary worksheet: fill in the blanks

This chapter describes several ways in which cells rearrange DNA to alter gene expression. Yeast cells rearrange DNA to change their .... Trypanosomes can alter their surface coat by expressing a different .... gene. This is also accomplished by DNA rearrangement, and allows the parasite to escape the host .... response. Yeast mating type switching and trypanosome surface antigen variation are similar in several ways. Both systems have multiple .... alleles of the rearrangeable gene but just one allele that is .... The expressed allele is located in one particular .... The allele present in the expressed locus can be changed when one of the .... alleles is .... into the expressed locus. This process requires a ....-............... .... in the chromosome at the expressed locus. In these DNA rearrangements, the donor unexpressed allele (**is / is not**) retained at its original location.

Another type of DNA rearrangement occurs in the transfer of .... from the bacterium .... to plant cells. The T-DNA causes growth of a plant .... that secretes ...., which the bacterium uses for .... The .... is copied from the .... in the bacterium and transferred to the plant cell, where it .... into the plant genome. This rearrangement does not involve a double-strand break; instead, a .... strand of the T-DNA is excised by making a .... at each end.

DNA rearrangement occurs in mammalian cells when they are exposed to the drug .... Cells respond to this drug by .... the DHFR gene. In .... amplification, the new copies of the DHFR gene are located at their .... position. .... amplification occurs when the new copies of the DHFR gene form extrachromosomal arrays called ....-............... ....

DNA rearrangement can be exploited to introduce exogenous DNA into eukaryotic cells. New genes can be introduced into plant cells by cloning them into .... and allowing *Agrobacterium* to transfer them into the plant genome. Several techniques can be used to introduce DNA into mammalian cells in order to make .... animals.

Additional reading

Herskowitz, I., Rine, J., and Strathern, J. (1992). Mating type determination and mating type interconversion in *Saccharomyces cerevisiae*. In *The molecular and cellular biology of Saccharomyces*, vol. 2 (E. Jones, J. Pringle, and J. Broach, eds) Cold Spring Harbor Laboratory Press, Cold Spring Harbor, New York, pp. 583–656.

Vanhamme, L. (1995). Control of gene expression in trypanosomes. *Microbiol. Reviews* **59**, 223–240.

Zupan, J. and Zambryski, P. (1995). Transfer of T-DNA from Agrobacterium to the plant cell. *Plant Physiol.* **107**, 1041–1047.

Chapter 33: Immune diversity

Multiple-choice questions

(1) **The humoral immune response involves:**
- (a) stimulation of B cells by helper T cells.
- (b) secretion of antibodies that specifically recognize foreign molecules.
- (c) presentation of an antigen–antibody complex to cytotoxic T cells.
- (d) destruction of the antigen–antibody complex by macrophages.
- (e) induction of the complement cascade to lyse target cells.

(2) **The cell-mediated immune response:**
- (a) is dependent upon the humoral response.
- (b) is usually used to respond to virus-infected cells.
- (c) involves direct recognition of the antigen by killer T cells.
- (d) requires that the antigen be presented to killer T cells by an MHC protein.
- (e) leads to target cell lysis.

(3) **An individual B cell produces many different antibodies.**
- (a) true   (b) false

(4) **An antibody molecule:**
- (a) is composed of two light and two heavy chains.
- (b) is held together by disulfide bonds between the chains.
- (c) binds antigen through its constant region.
- (d) has certain effector functions as determined by its light chain.
- (e) is a W-shaped molecule.

(5) **Which of the following statement(s) about immunoglobulin gene rearrangements is/are true?**
- (a) All species have the same number of V genes.
- (b) The J segment is part of the constant region.
- (c) Deletions and rearrangements can be generated when segments are joined.
- (d) Once a productive rearrangement has occurred at one allele, the other allele also undergoes rearrangement.

(6) **Which of the following statement(s) about class switching is/are true?**
- (a) Each class of heavy chain has a different function.
- (b) Class switching follows the order of the heavy chain genes in the chromosome.
- (c) Once a class switch has taken place, no further switches are possible.
- (d) The heavy chain can also be changed by alternative splicing.

(7) The T-cell receptor:

(a) is required for helper T cells to stimulate B cells in the humoral response.

(b) is required for killer T cells to recognize antigen.

(c) recognizes antigen fragments and MHC molecules independently.

(d) has a domain organization that is roughly similar to that of immunoglobulin molecules.

(e) genes undergo rearrangement analogous to that of immunoglobulin genes.

Concept questions

(1) Except in cases of autoimmune disease, the immune system does not react against 'self'. How is this achieved?

(2) Describe the steps in heavy chain immunoglobulin gene rearrangement.

(3) List the ways in which immune diversity is generated.

(4) Give an example of how cells of the immune system interact. What molecules are involved?

(5) What is the fate of a J segment on the 5' side of the J segment that is recombined with the V segment? What is the fate of a J segment on the 3' side of the one that is recombined with the V segment?

Problems

(1) Suppose that immunoglobulin genes were encoded intact in the genome instead of being assembled from pieces. Assuming that a human can respond to one million different antigens, how many genes would need to be devoted to the immune response? What fraction of the human genome would these genes occupy?

(2) Rearrangement of V, D and J segments is governed by consensus sequences found next to each segment. Two types of consensus sequences, the heptamer and the nonamer, are found in different orientations and with different spacing between them. Recombination can occur only between segments with opposite orientation and different spacing. For example, Vκ genes are followed by a heptamer and a nonamer with 12 nucleotides between them. Every Jκ is preceded by a nonamer and heptamer separated by 23 nucleotides. Therefore, Vκ can be joined to Jκ.

(a) What would happen if you changed the spacing and orientation of the sequences following Vκ?

(b) What would happen if you also changed the spacing and orientation in front of Jκ?

(c) At the heavy chain locus, what would happen if you changed the spacing and orientation following V?

(3) (a) Joining of immunoglobulin gene segments is initiated by a double-stranded break at the heptamer site. This requires the RAG1 and RAG2 proteins. What happens to mice with mutations in the genes encoding RAG1 or RAG2?

(b) The coding ends, which will be joined together, first undergo a hairpin reaction that joins the 3' end of the top strand to the 5' end of the bottom strand (refer to Figure 33.14). The top strand is then nicked to create a single-strand overhang on the 3' end. This overhang is filled in by insertion of nucleotides (these are the P nucleotides) to create a blunt-ended double-stranded fragment. What kind of enzyme would be required to fill in the single-strand overhang?

(4) Surface antigen variation and mammalian immune diversity both result from DNA rearrangements. Trypanosomes rearrange DNA to express one of the approximately one thousand different VSG genes they carry. The mammalian immune system can rearrange DNA to create hundreds of thousands of different antibodies, including antibodies that react with any VSG protein. Despite this numerical superiority, the trypanosomes successfully evade the host immune system. How is this possible?

Summary worksheet: fill in the blanks

The immune response relies on B and T ...., which produce proteins that specifically bind to foreign antigens. .... cells make proteins called .... These proteins are composed of .... heavy chains and .... light chains, which have both .... and .... regions. Genes encoding intact immunoglobulin proteins (**are / are not**) found in the germline; rather, they are .... from gene .... By using different gene segments in different combinations, the immune system generates an enormous .... of immunoglobulin proteins without requiring an enormous number of genes. Light chain genes are assembled from three segments known as ...., ... and .... Heavy chain genes contain an additional .... segment, located between the .... and the .... segments.

The rearrangement of immunoglobulin genes is governed by the .... and .... consensus sequences found at the ends of all gene segments that can recombine. During rearrangement, the DNA is cut in the .... consensus sequence and joined to another .... Additional nucleotides called .... and ...., which are not encoded in the genome, can be added at the junction to cause further alterations in the coding region.

180

Diversity of immunoglobulins results from several sources: different combinations of .... and .... gene segments (in addition to these, heavy chains also include .... segments), different nucleotides at each segment ...., different combinations of .... and .... chains, and .... mutation of the assembled immunoglobulin gene during the life of the B cell.

Not all immunoglobulin gene rearrangements are productive; some can generate genes that are incomplete or that change reading .... Once a productive rearrangement has occurred, further rearrangements (**are / are not**) prohibited on both copies of that locus (one on each copy of the chromosome). This is known as ...., and ensures that each B cell makes only .... immunoglobulin molecule.

Although the V-D-J junction of heavy chain genes does not change after a productive rearrangement has occurred, the .... region can change through the process of .... The different classes of C regions impart different .... functions, such as membrane anchoring or complement activation, to the immunoglobulin molecule without changing its antigen specificity.

T-cell receptors are similar to immunoglobulin molecules except that they contain .... chains. T-cell receptor genes are assembled from gene segments in a process that (**is / is not**) similar to immunoglobulin gene rearrangement. The T-cell receptor only recognizes its antigen when the antigen is bound by an .... protein. Other proteins encoded at the MHC locus include the .... proteins, which .... cells bearing foreign antigens.

Additional reading

Bjorkman, P. (1997). MHC restriction in three dimensions: a view of T cell receptor/ligand interactions. *Cell* **89**, 167–170.

Bogue, M. and Roth, D. (1996). Mechanism of V(D)J recombination. *Curr. Opin. Immunol.* **8**, 175–180.

Lewis, S. and Wu, G. (1997). The origin of V(D)J recombination. *Cell* **88**, 159–162.

Rajewsky, K. (1996). Clonal selection and learning in the antibody system. *Nature* **381**, 751–758.

Ramsden, D., van Gent, D., and Gellert, M. (1997). Specificity in V(D)J recombination: new lessons from biochemistry and genetics. *Curr. Opin. Immunol.* **9**, 114–120.

Part 7: Cell growth, cancer, and development

Theme

In the first six parts of *Genes VI*, you learned about many steps in gene expression and the many stages at which gene expression can be regulated. In Part 7, you will see this regulation in action at a higher level. You will also find many examples of gene products (usually proteins) acting to influence the expression of other genes. Throughout Part 7, a central theme of protein activation will emerge: proteins are phosphorylated by specific kinases, this phosphorylation causes the protein to change conformation, this conformational change allows the protein to interact with effector proteins.

We have described the ways in which gene activity can be affected by chromatin structure, transcriptional regulation, and processing and transport of the mRNA. However, gene expression is not completed until the protein product has been synthesized, modified and delivered to its proper site of action. Chapter 34 deals with how proteins mature inside the cell, how they are secreted and how they are directed to their intended target sites.

In Chapter 35 you will learn how cells receive signals from their environment or from other cells. A wide variety of signal transduction pathways transmit these signals to the nucleus, where most of these signals affect gene activity. Here, we will find an abundance of protein–protein interactions with several unifying themes. In almost every signal transduction pathway, proteins are phosphorylated on specific amino acids. This phosphorylation alters the proteins' properties, for example by causing a conformational change or by stimulation of an enzymatic activity. These changes in turn allow interaction with target proteins, which themselves are often phosphorylated, thus transmitting the original signal further down the pathway.

We have already discussed examples of proteins whose functions are altered by phosphorylation. Examples include histone H1 (Chapter 27) and the C-terminal domain of RNA polymerase II (Chapter 28). In addition, we have described many processes, from DNA replication to mRNA polyadenylation, that rely on specific interactions between proteins. However, signal transduction pathways such as the Ras pathway rival any other process in the extensive involvement of protein phosphorylation and protein–protein interactions.

Protein phosphorylation is also a major means of controlling the cell division cycle (Chapter 36). For example, cells enter mitosis when M phase kinase phosphorylates its target proteins. The catalytic activity of M phase kinase is itself regulated by phosphorylation. Another interesting mechanism used in cell cycle regulation is protein degradation. One subunit of M phase kinase is a cyclin protein, which is degraded during M phase, leading to inactivation of the kinase.

Chapter 37 merges the concepts presented in Chapters 35 and 36. You will see how alterations in signal transduction pathways activate cell division inappropriately, leading to oncogenesis. Virtually any step of a signal transduction pathway can go awry, causing uncontrolled cell division.

Part 7 contains many examples of how eukaryotes carefully choreograph the activation of their genes. This choreography reaches its pinnacle in the process whereby a fertilized egg develops into a differentiated organism. Chapter 38 provides an introduction to the roles of gene activity in this development. This chapter describes some of the elegant ways in which organisms activate genes in a spatially restricted manner, allowing one genome to create an organism containing many different types of cells.

Key terms

Use the following terms as a guide for your study and for the completion of the summary at the end of each chapter.

Chapter 34

❑ ARF	❑ exocytosis	❑ preformed oligosaccharide
❑ cisternae	❑ glycosylation	
❑ clathrin	❑ Golgi	❑ protein sorting
❑ coated pits	❑ ligand	❑ receptor
❑ coatomer	❑ lumen	❑ receptor-mediated endocytosis
❑ complex oligosaccharide	❑ N-linked glycosylation	
❑ dolichol	❑ NSF	❑ SNAP
❑ endocytosis	❑ O-linked glycosylation	❑ SNARE
❑ endoplasmic reticulum		❑ vesicle

183

Chapter 35

- ❑ active transport
- ❑ autophosphorylation
- ❑ channel
- ❑ CREB
- ❑ effector
- ❑ G protein
- ❑ ion channel
- ❑ ion gradient
- ❑ JAK-STAT
- ❑ MAP kinase
- ❑ passive transport
- ❑ protein serine/ threonine kinase
- ❑ protein tyrosine kinase
- ❑ ras
- ❑ receptor
- ❑ second messenger
- ❑ SH2 domain
- ❑ SH3 domain
- ❑ transporter
- ❑ voltage-gated channel

Chapter 36

- ❑ apoptosis
- ❑ *cdc* mutant
- ❑ cdc2
- ❑ CDC28
- ❑ cell cycle
- ❑ control points
- ❑ cyclin
- ❑ Fas
- ❑ G$_0$ phase
- ❑ G$_1$ phase
- ❑ G$_2$ phase
- ❑ M phase kinase
- ❑ maturation promoting factor
- ❑ mitosis
- ❑ p34
- ❑ RB
- ❑ S phase
- ❑ START

Chapter 37

- ❑ amplification
- ❑ *bcr–abl*
- ❑ c-*onc*
- ❑ constitutive mutation
- ❑ growth factor
- ❑ growth factor receptor
- ❑ insertion
- ❑ Myc
- ❑ oncogene
- ❑ p53
- ❑ promoter activation
- ❑ promoter insertion
- ❑ proto-oncogene
- ❑ Ras
- ❑ Rous sarcoma virus
- ❑ RB
- ❑ Src
- ❑ transformed cell
- ❑ transforming virus
- ❑ translocation
- ❑ tumor suppressor gene
- ❑ v-*onc*

Chapter 38

- ❑ *ANT-C* complex
- ❑ axis
- ❑ *bicoid*
- ❑ *BX-C* complex
- ❑ gap gene
- ❑ gradient
- ❑ homeobox
- ❑ homeotic gene
- ❑ *Hox* gene
- ❑ *hunchback*
- ❑ localization
- ❑ maternal-effect gene
- ❑ pair-rule gene
- ❑ parasegment
- ❑ polarity
- ❑ segment
- ❑ segmentation gene
- ❑ segment polarity gene
- ❑ syncytial blastoderm

184

Chapter 34: Protein trafficking

Multiple-choice questions

(1) **Which statements about the glycosylation of secreted proteins are true?**

(a) Almost all secreted proteins are glycosylated.

(b) Glycosylation occurs exclusively on the NH_2 group of asparagine.

(c) An oligosaccharide is formed on the lipid dolichol and then transferred to the protein.

(d) The oligosaccharide is trimmed by the removal of mannose residues.

(e) Processing of the oligosaccharide is complete before the protein leaves the ER.

(2) **The Golgi:**

(a) contains a series of stacked internal cisternae.

(b) is a polar organelle with its *cis* face close to the ER and its *trans* face close to the plasma membrane.

(c) contains the same enzymes in all its parts.

(d) is the site of further modification of preformed oligosaccharides to make complex oligosaccharides.

(3) **Proteins are transported around the cell in membranous vesicles. These vesicles:**

(a) form when a section of membrane protrudes and buds off.

(b) have a layer of coat protein around the inside of the vesicle.

(c) use their protein coat to find their target membrane.

(d) are uncoated after they reach their target.

(4) **Clathrin:**

(a) is a component of all endocytic and secretory vesicles.

(b) forms a triskelion composed of three heavy chains and three light chains.

(c) forms a network on the outside of coated vesicles.

(d) binds to an inner shell of adaptor proteins.

(e) is responsible for targeting vesicles to their correct target membranes.

(5) **Which statements about receptor-mediated endocytosis are true?**

(a) Receptors do not internalize until they bind their ligands.

(b) Receptors can diffuse through the membrane in two dimensions into coated pits.

(c) After forming a vesicle, receptors are targeted directly to the lysosome.

(d) The cytoplasmic region of the receptor interacts with adaptor proteins.

Concept questions

(1) Describe how secreted proteins fold.

(2) List the steps in the maturation of a secreted protein.

(3) What molecule initiates the budding of clathrin-coated vesicles? How does this process work?

(4) What molecules are involved in vesicle fusion?

(5) Describe the possible fates of a receptor–ligand complex.

Summary worksheet: fill in the blanks

Secreted proteins undergo posttranslational modification by the addition of .... Proteins are modified as they travel through the .... and the .... When moving to a new location (either another organelle or to the cell membrane), proteins are carried in membranous .... that are surrounded by protein .... Most of these vesicles contain the protein ...., which is important for the structure and function of the vesicle. These .... also contain specific proteins that direct the vesicle to its intended .... membrane. .... are cell surface proteins that bind specific .... and bring them into the cell in a process known as ....

Additional reading

Corsi, A. and Schekman, R. (1996). Mechanism of polypeptide translocation into the endoplasmic reticulum. *J. Biol. Chem.* **271**, 30299–30302.

Pelham, H. and Munro, S. (1993). Sorting of membrane proteins in the secretory pathway. *Cell* **75**, 603–605.

Robinson, M. (1994). The role of clathrin, adaptors, and dynamin in endocytosis. *Curr. Opin. Cell Biol.* **6**, 538–544.

Rothman, J. and Wieland, F. (1996). Protein sorting by transport vesicles. *Science* **272**, 227–234.

Schekman, R. and Orci, L. (1996). Coat proteins and vesicle budding. *Science* **271**, 1526–1533.

Takizawa, P. and Malhotra, V. (1993). Coatomers and SNAREs in promoting membrane traffic. *Cell* **75**, 593–596.

Chapter 35: Signal transduction

Multiple-choice questions

(1) **Signal transduction describes the ways in which external molecules cause internal changes in cells. These processes:**
   (a) always involve the movement of ligands across the membrane.
   (b) amplify the strength of the external stimulus.
   (c) rely on cell surface receptors that can interact with intracellular proteins.
   (d) can trigger the production of a second messenger inside the cell.

(2) **The concentrations of various ions in the cell are carefully maintained at specific levels. Mark all statements about this process that are true.**
   (a) Charged molecules can diffuse freely across a lipid membrane.
   (b) Cells contain more free anions than cations.
   (c) Each ion has a specific electrochemical gradient across the cell membrane.
   (d) When the extracellular concentration of an ion is lower than the intracellular concentration, the ion can enter the cell by passive transport.

(3) **Receptor tyrosine kinases:**
   (a) have a cytoplasmic kinase domain.
   (b) dimerize upon ligand binding.
   (c) change conformation of their cytoplasmic domains upon ligand binding.
   (d) become catalytically active after the conformational change.
   (e) autophosphorylate upon becoming catalytically active.

(4) **SH2 domains:**
   (a) are found in a wide variety of signaling proteins.
   (b) are approximately 100 amino acids long.
   (c) bind to tyrosine kinases after they autophosphorylate.
   (d) bind all SH2-binding domains with equal affinity.
   (e) are usually found in proteins that also have catalytic activity.

(5) **The Ras protein:**
   (a) binds directly to extracellular ligands.
   (b) is activated when the adaptor protein Grb2 binds the SOS protein.
   (c) binds GDP in its inactive state.
   (d) exchanges GDP for GTP when activated by SOS.
   (e) activates the Raf kinase, which in turn activates a MEK kinase.

(6) **MAP kinases:**

    (a) are activated by the Ras pathway.

    (b) phosphorylate their targets on tyrosine residues.

    (c) act to stimulate or repress the transcription of target genes through the phosphorylation of transcription factors.

    (d) become active when phosphorylated by MEK kinases.

Concept questions

(1)   How do carrier proteins and ion channels differ? In what ways are they similar?

(2)   List the steps in G protein signal transduction.

(3)   Where are receptor protein kinases found? What kind of ligands do they bind? What are their substrate specificities? What effector pathways do they activate?

(4)   What are the effects of autophosphorylation on protein tyrosine kinases?

(5)   Platelet derived growth factor (PDGF) can activate the Elk-1 transcription factor. What molecules are involved in this activation?

(6)   Give an example of how the activity of monomeric G proteins is controlled by other proteins.

(7)   List three ways in which a MAP kinase can transport a signal across the nuclear envelope.

(8)   How are trimeric G proteins activated? How are monomeric G proteins activated?

(9)   Mutations in the yeast *STE5* gene affect signal transduction at several levels. Propose an explanation.

(10) How do JAK kinases affect transcription? How does this pathway compare with activation by monomeric G proteins?

Summary worksheet: fill in the blanks

Cells need to respond to external ...., either from their environment or from other cells. Many types of signaling molecules do not enter the cell. Rather, they interact with specific .... found on the cell surface. These .... activate a variety of .... pathways that ultimately transmit the signal to the ...., where gene expression is altered.

One common pathway uses trimeric ...., so named because they bind .... These proteins interact with

receptors when the receptors are in their .... state but
.... when the receptor binds its .... After dissociation,
the .... subunit acts on its target to stimulate the production of a
.... messenger such as ....

Another pathway uses tyrosine kinase receptors, which ....
upon ligand binding. This activates a .... domain present in the
.... portion of the receptor. Some receptors activate the ras
pathway. Ras is a .... G protein and (**does / does not**) bind
directly to the receptor. It is activated when the .... protein
causes it to exchange .... for .... Ras activates the
.... kinase, which in turn can activate a .... pathway.
The end result of these pathways is alteration in the pattern of
....

Additional reading

Bourne, H. (1997). How receptors talk to trimeric G proteins. *Curr. Opin. Cell Biol.*
9, 134–142.

Cohen, G., Ren, R., and Baltimore, D. (1995). Modular binding domains in signal
transduction proteins. *Cell* **80**, 237–248.

Herskowitz, I. (1995). MAP kinase pathways in yeast: for mating and more. *Cell* **80**,
187–197.

Hill, C. and Treisman, R. (1995). Transcriptional regulation by extracellular signals:
mechanisms and specificity. *Cell* **80**, 199–211.

Hunter, T. (1995). Protein kinases and phosphatases: the yin and yang of protein
phosphorylation and signaling. *Cell* **80**, 225–236.

Lefkowitz, R. (1993). G protein-coupled receptor kinases. *Cell* **74**, 409–412.

Treisman, R. (1996). Regulation of transcription by MAP kinase cascades. *Curr.
Opin. Cell Biol.* **8**, 205–215.

Chapter 36: Cell cycle and growth regulation

Multiple-choice questions

(1) **M phase kinase:**
  (a) can cause G_2-blocked *Xenopus laevis* oocytes to enter M phase.
  (b) contains a kinase subunit called p34.
  (c) contains a regulatory subunit that is a cyclin.
  (d) phosphorylates target proteins at specific times in the cell cycle.

(2) **Which statements about cyclins are true?**
  (a) There are several families of cyclins that differ in amino acid sequence and in their properties.
  (b) Cyclins are present at a constant steady-state level throughout the cell cycle.
  (c) Cyclins are degraded rapidly when the cell enters mitosis.
  (d) Cyclins regulate the activity of the M phase kinase.

(3) **In *Schizosaccharomyces pombe*, cdc2:**
  (a) is the homolog of the *Xenopus laevis* p34 protein.
  (b) forms a complex in M phase with the cyclin cdc13.
  (c) is controlled by phosphorylation.
  (d) has no homolog in *Saccharomyces cerevisiae*.

(4) **Which of the following does *not* occur when a cell enters M phase?**
  (a) Chromatin condenses.
  (b) Histone H1 is dephosphorylated.
  (c) The nuclear envelope, the endoplasmic reticulum and the Golgi break down.
  (d) The spindle is formed.
  (e) The contractile ring is formed.

(5) **Apoptosis:**
  (a) is the programmed death of specific cells.
  (b) is a normal process during development.
  (c) can only be induced when the Fas or TNF receptors bind their ligands.
  (d) involves condensation of the nucleus, DNA fragmentation, protein degradation and the destruction of membranes.

Concept questions

(1) At what stages in the cell cycle are checkpoints found? What is 'checked' at each point? What would cause a cell to fail at a checkpoint?

(2) How is p34 kinase activated? How is it inactivated?

(3) How does the phosphorylation of cdc2 control the cell cycle in *S. pombe*?

(4) What is the role of the RB protein in G_1?

(5) What evidence demonstrates that the chromosome cycle, the centrosome cycle and the cytoplasmic cycle are all under coordinate control? What evidence demonstrates that they can proceed independently?

Summary worksheet: fill in the blanks

The cell cycle is controlled at several .... occurs in G_1 phase and is the point at which the cell becomes committed to .... Another checkpoint occurs in G_2 phase before the cell becomes committed to ....

Mitosis is initiated by the ...., which contains .... subunits. One subunit, called .... (the *S. pombe* homolog is called .... and the *S. cerevisiae* homolog is known as ....), is a .... that phosphorylates specific target proteins. This subunit is regulated by .... To be active, two amino acids within an ....-binding domain must be dephosphorylated and one other amino acid must be phosphorylated. The other subunit is a .... that also regulates the activity of the catalytic subunit. These .... accumulate throughout the cell cycle but are rapidly .... during .... phase. Without the .... subunit, the .... subunit is inactive.

The transition from G_1 to .... phase is controlled by a kinase that contains the same .... subunit but a different .... subunit. Numerous other proteins, originally identified as .... (which stands for ....) mutants, are required for the cell cycle. These proteins function to duplicate and partition the .... and .... Upon entry into .... phase, the cell is dramatically .... This process includes .... of chromatin, breakdown of the ...., ... and ...., and reorganization of microtubules into the ....

191

Additional reading

Jacobson, M., Weil, M., and Raff, M. (1997). Programmed cell death in animals. *Cell* **88**, 347–354.

Lew, D. and Kornbluth, S. (1996) Regulatory roles of cyclin dependent kinase phosphorylation in cell cycle control. *Curr. Opin. Cell Biol.* **8**, 795–804.

Nurse, P. (1994). Ordering S phase and M phase in the cell cycle. *Cell* **79**, 547–550.

Paulovich, A., Toczyski, D., and Hartwell, L. (1997). When checkpoints fail. *Cell* **88**, 315–321.

Sherr, C. (1993). Mammalian G1 cyclins. *Cell* **73**, 1059–1065.

Williams, G. and Smith, C. (1993). Molecular regulation of apoptosis: genetic controls on cell death. *Cell* **74**, 777–779.

Chapter 37: Oncogenes and cancer

Multiple-choice questions

(1) **Transformed cells can be cultured indefinitely, while normal cells will divide only a finite number of times in culture.**

(a) true   (b) false

(2) **Mutations in the *ras* gene are often associated with oncogenesis. These *ras* mutations:**

(a) can occur in v-*ras* or c-*ras*.

(b) generally alter a single amino acid such as glycine-12 or glutamine-61.

(c) can cause a constitutively active Ras that does not hydrolyze GTP and that cannot be inactivated by GAP.

(d) sometimes cause oncogenesis by overexpressing the wild-type Ras protein.

(3) **Chromosome translocations can cause oncogenesis by:**

(a) bringing an *onc* gene near an actively expressed locus such as an Ig locus.

(b) changing the structure of an *onc* gene.

(c) bringing an *onc* gene into a region of constitutive heterochromatin.

(d) creating a hybrid gene by fusing two coding regions, as in *bcr–abl*.

(4) **Src is a membrane-associated tyrosine kinase that can become oncogenic. Which statements about Src are true?**

(a) The Src protein contains SH2, SH3, catalytic and suppressor domains.

(b) c-Src has higher kinase activity than v-Src.

(c) Autophosphorylation of tyrosine-416 is associated with higher kinase activity.

(d) Phosphorylation of tyrosine-527 is associated with higher kinase activity.

(e) Tyrosine-527 is missing from v-Src.

(f) Phosphorylation of tyrosine-527 could allow the C-terminal suppressor domain to interact with the N-terminal SH2 domain, keeping Src inactive. When tyrosine-527 is not phosphorylated, the SH2 domain is free to interact with receptors.

(5) **The RB protein:**

(a) induces S phase by binding to the E2F transcription factor.

(b) is phosphorylated by a cyclin/cdk complex.

(c) releases E2F when it is phosphorylated.

(d) can be sequestered by certain viral proteins called tumor antigens, allowing E2F to function.

Concept questions

(1) Oncogenes carried by transforming retroviruses cause oncogenesis, but the same genes present at their normal chromosomal locations do not. Why not?

(2) Retroviruses can cause oncogenic transformation even if they do not carry an *onc* gene. List three ways in which this can happen.

(3) Many oncogenes encode proteins that participate in signal transduction pathways. Which types of signal transduction proteins can become oncogenic? Give an example of each, and suggest how it might lead to uncontrolled cell division.

(4) In what ways can mutant transcription factors cause transformation?

(5) What changes in the epidermal growth factor receptor (Erb) lead to oncogenesis?

(6) What functional domains are found in the p53 protein? Which parts of p53 are mutated in tumor cells?

Summary worksheet: fill in the blanks

Oncogenesis can result from the inappropriate expression of .... genes, which causes cells to divide when they should not. Retroviruses that acquire cellular genes (.................... genes) can cause oncogenesis by .... of the wild-type gene. Other .... viruses can be oncogenic by expressing a .... form of the gene. Even retroviruses that do not carry a .... gene can be oncogenic if they .... into the host chromosome near a .... gene and alter its .... Onc genes can also be created by chromosome ....

Many *onc* genes encode proteins that function in .... pathways. Proteins at any point in the pathway can become oncogenic including .... factors, ... factor .... , G proteins (...................., for example), intracellular .... and other signaling molecules. These factors become oncogenic when they .... to a form that (**does / does not**) respond to their normal control mechanisms. Ultimately, oncogenes cause inappropriate .... In fact, many .... factors themselves can mutate to oncogenic forms.

The tumor suppressor .... functions by sequestering the transcription factor ...., which activates genes required for

194

.... phase. The tumor suppressor .... can stop the cell cycle in .... (for example, if DNA is ....). This protein can also trigger ....

Additional reading

Baserga, R. (1994). Oncogenes and the strategy of growth factors. *Cell* **79**, 927–930.

Hunter, T. (1997). Oncoprotein networks. *Cell* **88**, 333–346.

Levine, A. (1977). p53, the cellular gatekeeper for growth and division. *Cell* **88**, 323–331.

Picksley, S. and Lane, D. (1994). p53 and Rb: their cellular roles. *Curr. Opin. Cell Biol.* **6**, 853–858.

Sherr, C. (1996). Cancer cell cycles. *Science* **274**, 1672–1677.

Weinberg, R. (1995). The retinoblastoma protein and cell cycle control. *Cell* **81**, 323–330.

Chapter 38: Gradients and cascades

Multiple-choice questions

(1) **Which statements about maternal effect genes are true?**
   (a) Mutant alleles have no effect on phenotype in the mother.
   (b) They encode gene products that are packaged into the egg.
   (c) They alone are sufficient to determine the developmental fate of the entire embryo.

(2) **Morphogens are spread uniformly throughout an unfertilized egg.**
   (a) true   (b) false

(3) *Bicoid* **mRNA:**
   (a) is required for anterior development.
   (b) is synthesized in nurse cells and transported into the egg.
   (c) is localized to the posterior part of the egg.
   (d) is found in a concentration gradient from anterior to posterior.
   (e) is translated before the egg is fertilized.
   (f) encodes a transcription factor that regulates *hunchback*.

(4) **Which statements accurately describe the hierarchy of genes involved in *Drosophila* development?**
   (a) Maternal effect genes establish gradients across the entire embryo.
   (b) These gradients activate or repress the gap genes.
   (c) Gap genes establish gradients across the entire embryo.
   (d) Gap genes activate or repress the pair-rule genes.
   (e) Pair-rule genes activate or repress the segment polarity genes.

(5) **Homeotic genes:**
   (a) define how each segment ultimately differentiates.
   (b) when mutant, can cause segments to transform into other segments.
   (c) are encoded by complex loci.
   (d) encode growth factor receptors.

(6) **Complex loci such as *ANT-C* and *BX-C* have many unusual features including:**
   (a) alternative promoters separated by a great distance
   (b) enormous introns
   (c) a genetic map that correlates with body plan
   (d) many *cis*-acting regulatory elements
   (e) many different protein products
   (f) a complicated pattern of intra-locus genetic complementation

Concept questions

(1)   In *Drosophila melanogaster*, three sets of maternal effect genes are responsible for the formation of the anterior–posterior axis. What are these gene systems? How do they work?

(2)   The Bicoid protein is a transcription factor present in a concentration gradient from anterior to posterior. Bicoid activates the transcription of *hunchback*. However, *hunchback* is expressed in a simple all-or-none fashion, not a gradient resembling *bicoid*. Why do the expression patterns of *bicoid* and *hunchback* differ?

(3)   How is the dorsal–ventral polarity of a *Drosophila* oocyte established?

(4)   What are the consequences of mutations in (a) gap genes; (b) pair-rule genes; (c) segment polarity genes?

(5)   What are the consequences of loss-of-function mutations in (a) homeobox genes; (b) gain-of-function mutations?

(6)   How are homeobox-containing genes of *Drosophila* and mouse *Hox* genes similar? How do they differ?

Summary worksheet: fill in the blanks

Development of a morphologically complex organism from a fertilized egg requires the action of many sets of genes. This process has been studied extensively in *Drosophila melanogaster*. .... genes are expressed in the mother and packaged into the egg. These gene products establish the .... and .... axes of the egg. For example, the *bicoid* mRNA is present in a concentration .... in the egg. In places where the bicoid concentration is above a certain threshold, it activates the *hunchback* gene, leading to the development of .... structures.

.... genes are expressed after fertilization. These are found in three types known as .... genes, ... genes and .... genes. They are expressed in patterns that divide the developing embryo into segments.

Finally, ... genes determine the ultimate development of each segment. These are encoded by .... loci such as .... and .... All these sets of genes work together in a combinatorial fashion to give each part of the embryo a specific developmental fate.

Additional reading

Curtis, D., Lehmann, R., and Zamore, P. (1995). Translational regulation in development. *Cell* **81**, 171–178.

Krumlauf, R. (1994). Hox genes in vertebrate development. *Cell* **78**, 191–201.

Lawrence, P. and Morata, G. (1994) Homeobox genes: their function in Drosophila segmentation and pattern formation. *Cell* **78**, 181–189.

Nüsslein-Volhard, C. (1996). Gradients that organize embryo development. *Sci. Am.* **275**, 54–61.

Schüpbach, T. and Roth, S. (1994). Dorsoventral patterning in Drosophila oogenesis. *Curr. Opin. Genet. Dev.* **4**, 502–507.

Chapter 1: Proteins

Answers to multiple-choice questions

1 (d,e); 2 (d); 3 (c); 4 (b); 5 (c); 6 (d,e); 7 (e); 8 (a); 9 (d); 10 (c).

Answers to concept questions

Question 1

macromolecule	subunit(s) and means of connection
polysaccharides	sugars connected by glycosidic bonds
lipids	fatty acids connected to glycerol by ester bonds
proteins	amino acids connected by peptide bonds
nucleic acids	nucleotides connected by 5'–3' phosphodiester bonds.

Question 2

The polypeptide chain as the protein's primary structure is organized in three dimensions based upon intra- and intermolecular interactions. A protein with sufficient hydrophobic residues can fold itself (or be folded by chaperones) so that all of the charged residues and most of the polar residues interact with one another while hydrophobic residues make up the periphery of the protein. In an aqueous environment, hydrophobic residues will be hidden from the water lattice while hydrophilic peptide moieties interact with the water lattice. Intramolecular interactions are responsible for the formation of β-sheet and α-helical structures and they contribute to the formation of active sites.

Question 3

To function as enzymes, proteins must be in an active conformation. This means that the protein fold must create an active site and this active site must be accessible to the substrate(s). Structural processes such as denaturation or misfolding generate inhibited enzymes. Kinetically, enzymes can be inhibited by covalent modifications such as (de)phosphorylation or (de)methylation, or by allosteric interactions, both of which eventually also affect enzyme structure. Further, feedback inhibition represents an additional regulatory opportunity in complex enzyme systems controlling inhibitor and/or substrate levels. Last but not least, enzymes can be inhibited from catalysis of a particular reaction by interactions with competitive ligands.

Answers to problems

Problem 1

High temperatures cause atoms to swing and hence weaken the bonds in molecules. While covalent bonds may withstand breaking, noncovalent bonds such as hydrogen bonds are most likely to be disrupted. Therefore, boiling of proteins causes breakage of their ternary and secondary structure, sometimes even degradation. RNAase is not degraded; however, the molecule loses its functional conformation. Your mistake was most likely that you cooled the RNAase sample down too quickly (e.g. on ice) so the protein remained in a nonfunctional conformation.

Problem 2

In order to make a cell contain a particular lipid that it does not naturally produce, one can genetically engineer the cell so that it (a) synthesizes a lipid uptake system (and provide the lipid in the growth medium) or (b) expresses an enzyme (complex) known to be able to utilize indigenous substrates and produce a lipid that is not naturally found in the cell. Lipids originate from glycerol and fatty acid synthesis and can be modified by inserting various moieties (sugars, amino acids or fatty acids) as the third side chain.

Answers to summary worksheet

This introductory chapter aims to remind you that of the four types of cellular macromolecules the **proteins** are the ones that take care of business' in the cell. Consisting of one or more **polypeptide** chains, the proteins are the direct expression product of the **genetic** information in the nucleic acids. While the nucleic acids function as the keeper of the secret of life, the **lipids** and **polysaccharides** are merely products of specific enzymic functions of proteins. In total, proteins have three distinct functions: i) they provide **structure** to the cell as architectural units, ii) they have **catalytic** activity which allows for the breakage and formation of bonds between atoms and molecules and iii) they **regulate** many cellular processes. The latter involves a *trans*-acting DNA binding protein function which is essential in the direct process of DNA perpetuation (Part 4) and transcription (Parts 3 and 6) and its regulation. Next, certain proteins keep others in a functional **conformation** or help them to mature and reach this functional **conformation**. Further, many proteins are involved in the activation and deactivation of other proteins usually as a result of minor modifications such as (de)phosphorylation or (de)methylation. In that capacity, many proteins contribute to **signal transduction** which is the initiation of an 'appropriate' cellular response to an external signal. In order to understand or even manipulate these processes, it is necessary to characterize at the molecular level the exact structure of the protein as well as the mode of action.

Chapter 2: Compartments

Answers to multiple-choice questions

1 (b,e); 2 (b,c,e); 3 (b); 4 (a,d); 5 (c); 6 (a); 7 (c,d); 8 (b); 9 (a); 10 (b).

Answers to concept questions

Question 1

The primary structure of a protein is the amino acid sequence of the polypeptide chain. A hydropathy analysis will identify the presence of hydrophobic regions suited for membrane incorporation. In addition, specific N-terminal and C-terminal amino acids allow the prediction of the final protein destination.

Question 2

Every cell is surrounded by a plasma membrane which is a lipid bilayer associated with transmembrane (= intrinsic) and peripheral proteins. Prokaryotic cells depend on the asymmetry of the plasma membrane in that all essential energy- and nutrient-harvesting processes can be assigned to the function of plasma membrane proteins. In eukaryotes, many of these functions have been relocated into organelles such as mitochondria and plant chloroplasts. While the genetic material ('nucleoid') resides freely in the prokaryotic cytoplasm, eukaryotes contain their genomic DNA in the nucleus (there is also organellar DNA). The perinuclear space extends into the cytosol as a gigantic lumen within the endoplasmic reticulum, which is the major location of protein processing. The Golgi system functions as another protein modification factory and a traffic switchboard. Both eukaryotic and prokaryotic cells contain many ribosomes; however, they are compartmentalized in eukaryotic cells.

Question 3

Biomembranes consist of different lipids and proteins. The two-leaflet lipid matrix is characterized by a very high lateral fluidity and a modest vertical exchange between the two leaflets. This is greatly affected by the degree of saturation of the fatty acid tails, the presence of 'stiffening' sterol-like compounds and enzymatic activity called 'flippase'. This flexible and fluid character allows for the incorporation of transmembrane and peripheral proteins, some of which are partially embedded in the membrane which allows for an asymmetric composition of membranes.

Question 4

Extracellular proteins in eukaryotic cells are exclusively synthesized by RER-associated ribosomes. The nascent polypeptides enter the RER lumen where they are processed predominantly by glycosylation. Vesicle transport allows entrance and exit of the Golgi system and facilitates exocytosis. This requires the fusion of the vesicles with the plasma membrane which changes the intracellular location of the protein into an extracellular one. Once across the plasma membrane, the protein is free to interact with the extracellular matrix.

Answers to problems

Problem 1

Association of proteins with the membrane can be accomplished in various ways: they can incorporate into the membrane, or interact through electrostatic forces with charged lipid headgroups or other membrane proteins, or crosslink via covalent bonds, but only the first and last of these interactions will lead to stable association with the membrane. Targeting to and incorporation into the membrane can be achieved by adding a signal peptide to the protein sequence; this is accomplished by translational (= in-frame) amino-terminal fusion. Covalent bonding may be achieved by expressing proteins that are able to modify (e.g., phosphorylate, aminate) and link the protein of interest to native membrane proteins (e.g, by esterization, peptidation).

Problem 2

Terminal differentiation means that cells stop proliferation once they reach a particular developmental stage in a tissue or organ. Hence, their cell cycle must be arrested in G_0, G_1 or G_2 phase. Tumerous cells lack this arrest and undergo practically unrestricted cell division cycling. Distinction is thus possible by monitoring the synthesis of certain kinases or cyclins that are exclusively produced during defined periods of the cell cycle. A general tumor therapy would be the disruption of the cell division cycle of these cells through mutagenesis (irradiation) or addition of inhibitors (chemotherapy).

Answers to summary worksheets

There are various ways of classifying all living organisms and structural, physiological, biochemical and molecular genetical criteria have all been used. They all support a fundamental distinction between prokaryotic and eukaryotic cells; however, there are common features to both which are what we know as the 'hallmarks of (cellular) life'. Both types of cells comprise a distinct compartment that is always separated from the chaotic environment by membranous **envelopes**. While prokaryotic cells can be considered a single compartment, a eukaryotic cell consists of many subcompartments including douple membrane-enveloped organelles, the endoplasmic reticulum and Golgi apparatus. Furthermore, inside and outside of the eukaryotic cell are structures collectively called the **cytoskeleton** that are involved in the organization of the cell's structure and function. From the standpoint of molecular genetics, these are not merely additional compartments and structures; instead, they contribute intimately to the complex processes of the maintenance and replication of genetic information as well as gene **expression** and protein **processing** in the cell. In that regard, it is crucial to understand cellular function at the macroscopic level (e.g., membrane fusion, microtubule assembly, nuclear breakdown or chromatin condensation). In addition, the understanding of both the prokaryotic and eukaryotic cell division cycles is crucial for the understanding of individual events and processes in the field of molecular genetics. While the karyotype is maintained in somatic cells through **mitosis, diploid** eukaryotic cells can reproduce also sexually. This requires gametogenesis which is implemented through **meiosis.** Meiosis and the fertilization process carry out the Mendelian segregation of traits.

Chapter 3: Genes are mutable units

Answers to multiple-choice questions

1 (e); 2 (a); 3 (c); 4 (e); 5 (c,e); 6 (a); 7 (c); 8 (e); 9 (c); 10 (a).

Answers to concept questions

Question 1
All visible and measurable characteristics of a cell or organism constitute its phenotype. The phenotype is ultimately determined by the genotype, the collection of all genetic information.

Question 2
A linkage group is the group of genetic loci that are inherited (segregated) together. It usually means that these loci reside very close to one another on the chromosome and that recombination (breakage) frequency in this region is low. Widely used examples are X-linked loci because of the distinct inheritance and phenotypic expression in male offspring only (e.g., 'white' mutation in *Drosophila*).

Question 3
Alleles are variants of homologous genes that usually reside on different chromosomes. In terms of mapping (RFLP or mutant phenotypic analysis), genes (the units of inheritance) are equivalent to loci on the chromosomes. These genetic terms have to be correlated with molecular terms that reflect the process of utilization of the genetic information and, hence, gene expression. The unit of gene expression is the transcriptional unit which may (usually in prokaryotes) or may not (in eukaryotes) contain many segments that can be translated into protein. A transcriptional unit that contains just one translatable unit is called a cistron. The cistron can be defined by complementation analysis: the failure of complementation in *trans* indicates that both mutations reside in the same transcriptional unit.

Question 4

A 'gain-of-function' mutation is a mutation that changes the product so that it remains fully functional but acquires an additional function. This could be the ability to use additional substrates which previously could not be used due to steric problems at the catalytic site.

Question 5

Polymorphism is the presence of multiple functional alleles in a population of a species that do not represent gain- or loss-of function mutations. Look at Figure 3.14 and note how the presence or absence of a four amino acid change in galactosyl transferase determines the final antigenicity of the O-antigen in blood.

Answers to problems

Problem 1

First of all, bacteria are haploid and complementation in *trans* requires the presence of a homologous functional gene. This has to be transferred into the bacterial cell, by transformation, conjugation or transduction, which leads to a merodiploid genotype. Gene, trait and chromosome are terms used to describe the linkage of genomic loci (units of inheritance) while cistron, transcriptional and translational units are units of function that relate to a physical map.

Problem 2

The segregation pattern after crossing between susceptible and resistant plants will show whether a loss of function is related to dominance. Then, differential mRNA analysis will identify unique differences between the susceptible and the resistant plant genomes. The unique cDNAs can then be inserted into an *Agrobacterium tumefaciens* Ti-plasmid for transformation of susceptible plants. This will constitute a transformation in *trans* and lead to a clone with the gene of interest. The gene can then be used to construct a molecular probe for isolation of the wild-type and mutant loci and to determine the nature of the mutation. This will eventually lead to a 'gene therapeutical' procedure designed to restore resistance.

Answers to summary worksheet

It has been known since the experiments of **Gregor Mendel's** that **traits** segregate to the progeny independently in a predictable way (following Mendel's laws). These traits are the visible results of the genetic activity of Mendel's 'discrete factors', the **genes,** which are the structural units of **inheritance**. Thus genes assort independently like the traits. Diploid cells contain at least two variants of a gene called **alleles** which are part of different **chromosome** homologs, hence **alleles** segregate following the pattern of chromosome distribution (in mitosis and meiosis). Genetic **loci** (or specific alleles) on a chromosome belong to a specific **linkage** group and are usually inherited together if they are physically close to one another. Chromosomes are not equal to **linkage** groups because of the crossing-over (= **reciprocal recombination**) during meiosis in which chromosome translocation occurs.

Each cell or multicellular organism has a defined and characteristic **genotype** which provides the blueprint for its morphological appearance, the **phenotype**. If this is the phenotype/genotype that has been first recognized for a particular species it is called the **wild-type**. Any deviation from the **wild-type** is considered a mutant phenotype/genotype. A mutation is the change in the sequence of DNA. If a population of a species contains more than **two** variants of a gene there is the problem that some species may be considered mutant phenotypes when compared with the wild-type phenotype. Because determination of the original variants of the species is difficult such a population 'genotype' is called **polymorphic**. This provides an evolutionary context to the change of genetic information by mutation in which **polymorphism** is resolved by fixation (creating a gene family) or loss. Both contribute to speciation and the presence of nonfunctional or pseudogenes in many recent genotypes.

While the gene was defined as the unit of **inheritance**, the **cistron** is the equivalent term for the unit of genetic function. The **cistron** codes for a single polypeptide chain that may be sufficient to constitute an enzyme or structural protein. It can be identified by **complementation in *trans*** of a mutation considering the ploidy (diploid cells usually have a mutant/wild-type genotype after mutagenesis) and genetic nature (eukaryotic or prokaryotic) of the cell. If mutations fail to complement in *trans* they are assigned to the same **complementation group**. You will learn in Parts 2 and 3 that prokaryotic transcripts are usually polycistronic. Consequently a(n insertional) mutation in the first cistron may affect transcription of the entire transcriptional unit and the complementation group will extend from a cistron to the entire **transcriptional unit**. In contrast, a eukaryotic cistron corresponds to a complementation group.

Chapter 4: DNA is the genetic material

Answers to multiple-choice questions

1 (c); 2 (b); 3 (c); 4 (a); 5 (b,d); 6 (b); 7 (a); 8 (c,d); 9 (b); 10 (b).

Answers to concept questions

Question 1
DNA is a simple polymer of nucleotides that are connected through phosphodiester bonds. Its double-stranded nature (secondary structure) allows faithful template-dependent synthesis. Utilization of the genetic

code allows this simple structure to encode the incredible diversity and complexity of all the peptides and proteins that we know.

Question 2

Some viruses have RNA as their genetic element; hence, DNA does not function as the 'transforming principle of inheritance'.

Question 3

Chemically, different nucleotides differ only in their nitrogenous bases. The information stored in DNA is the sequence of bases, and as these bases do not participate in the covalent link between the nucleotides, the information stored in the DNA does not affect the structure of the molecule and a change in this information (resulting from mutation or recombination) will not destroy the molecule.

Question 4

The percentage of 'G' is predictable only in double-stranded DNA because of the equal content of complementary bases in the molecule: %G = [G+C]/2. The [G+C] content can be determined spectrophotometrically.

Question 5

The process of mutagenesis occurs when a mutagen causes a mutation (change in the nucleotide sequence) changing the wild-type organism into a mutant. This change in genotype may manifest itself in a change in the phenotype (visually or quantitatively different characteristic). Point mutations are single base changes which can be a base substitution, insertion or deletion. If a given mutation is phenotypically reversed through an additional mutation, this is called a suppression.

Answers to summary worksheet

One of the most important and exciting discoveries in the history of science was that inheritance can be attributed to a single kind of molecule, the **nucleic acids**. The time between 1928 and 1952 was filled with many ingenious experiments that tried to prove that the transforming principle is **deoxyribonucleic acid**. The realization that the content of eukaryotic chromosomes, the transfecting principle in eukaryotic cells, the transforming principle in bacteria and the transduction of phage material can be reduced to the same molecule finally established that nucleic acid is the universal genetic material. Theoretically the major obstacle to accepting this fact was the very **simple structure** of nucleic acids which was known to be a **polynucleotide** chain. DNA, the genetic material in all prokaryotic and eukaryotic cells, exists as an **antiparallel** double strand of 2-deoxyribose molecules linked by 5'–3' **phosphodiester** bonds. The genetic information resides solely in the kind and sequence of **nitrogenous bases** that are linked to position 1 of the ribose by a **glycosidic** bond. The structure of the bases dictates that each of them pairs with only one other naturally: the purine **guanine** (G) pairs with the pyrimidine **cytosine** (C) and the purine **adenine** (A) pairs with the pyrimidine **thymine** (T).

Because a G•C pair establishes three **hydrogen** bonds G•C pairs are more stable than A•T pairs which involve only two hydrogen bonds. Bases that can pair are called **complementary** bases. This structure seemed to be too simple to contain all the information for the complicated structure of a cell. This obvious discrepancy was resolved by understanding two further mysteries: (i) how can DNA code for the incredible diversity of **protein** structures and (ii) how can the genetic information be maintained faithfully over generations? The solution to the first mystery was the finding that the genetic information stored in the bases and their sequence (= the **nucleotide** sequence) of selected segments of the DNA is 'translated' into a sequence of **amino acids** in a polypeptide following a universal code, the **genetic code**. This code reliably assigns **trinucleotides** to a specific amino acid. The polypetides in turn assemble to and/or function as structural proteins or enzymes both of which contribute to the maintenance and complexity of a cell. The second mystery was resolved by the discovery that the replication of DNA is **semiconservative** or that 'new' is always paired with 'old'. This means that daughter cells always start their independent cellular life with a complete genome consisting of a **parental** strand and a new complementary strand of DNA. By taking this into account it is a next logical step to explain variation between cells and organisms with distinct changes in the **nucleotide sequence**. These changes can be generated by the substitution of a single base called **point mutation** or by the **insertion** or **deletion** of one to thousands of nucleotides.

Chapter 5: Nucleic acid structure

Answers to multiple-choice questions

1 (c,d); 2 (b); 3 (a,c,e); 4 (a); 5 (a,d); 6 (b); 7 (a,c,e).

Answers to concept questions

Question 1
DNA renaturation (annealing) is facilitated by decreasing temperature and pH value and by increasing salt concentration.

Question 2
Stem–loop structures formed in a single-stranded nucleic acid such as RNA are intrinsic transcriptional terminators.

Question 3
A defined groove structure is necessary for DNA–protein interactions so that particular amino acids of DNA-binding and -modifying proteins can interact with particular bases.

Question 4

See Figure 1.5: a nucleoside consists of a pentose sugar and a nitrogenous base, whereas a nucleotide is a nucleoside with a tri-, di-, or monophosphate at the fifth position. DNA and RNA differ in that the pentose sugar in DNA lacks the hydroxyl group at the second position. With this additional functional group and the fact the RNA is usually single stranded, RNA is much less stable than DNA. mRNAs have half lives varying from minutes (in bacteria) to hours.

Answers to problems

Problem 1

Your assay probably did not work because the DNA was too highly coiled to allow effective transcription. Try adding DNA gyrase, which removes negative supercoils and transforms the DNA into a more relaxed state.

Problem 2

Melting and annealing of DNA are dependent on the relative content of G+C versus A+T, because the three hydrogen bonds of G•C pairs are more temperature stable than the two bonds between A•T. Therefore, G•C tends to associate more unspecifically than A•T. Primers with lower melting temperatures are thus less suited for the PCR than primers with higher T_m values.

Answers to summary worksheet

While the structures of polynucleotide chain duplexes are already fairly rigid, their organization in a **double helix** and the additional twisting of the duplex around its axis (**supercoiling**) may seem to 'freeze' the nucleic acid into a completely inflexible molecule. This, however, is not the case for several reasons: first, the structure of a double-stranded nucleic acid is strictly dependent on several environmental conditions including **temperature**, pH value and the concentration of salts (e.g. magnesium ions) and it will thus change with fluctuations in these parameters. Second, the cell has a toolkit of enzymes which modify structure and topology of DNA such as **topoisomerases**. Third, but not last, the DNA of a circular or linear chromosome/genome can fit into tight spaces and endure mechanical stresses. The complex structure of the DNA duplex molecule represents a successful adaptation that allows for faithful replication and the organized utilization of the genetic information. The ability of **complementary** segments within a **single-stranded** nucleic acid to form partial duplex structures accounts for **secondary** structures. These hairpin or **stem–loop** structures play a role in the regulation of DNA modification processes in that they influence the performance of polymerases and other DNA-binding proteins. In Parts 2 to 7 you will learn much about how particular nucleic acid structures influence the maintenance and retrieval of genetic information.

Chapter 6: Isolating the gene

Answers to multiple-choice questions

1 (c); 2 (b,d); 3 (b); 4 (c); 5 (a); 6 (a,e); 7 (b); 8 (a,c,e); 9 (b,d,e); 10 (a).

Answers to concept questions

Question 1

Endonucleases cut DNA at specific recognition sites thereby creating fragments of specific lengths. By using several different enzymes singly and in combination, it is possible to create a physical map of restriction sites. The basic feature of the gene that makes restriction analysis suited for mapping is its linear character.

Question 2

If the DNA of one organism has mutated so that a recognition site for a particular restriction enzyme is created (or destroyed), then when the DNA from that organism and another unmutated one is cut with that enzyme, the resulting fragments will be of different lengths in the two samples. This is called restriction fragment length polymorphism (RFLP).

Question 3

Sanger sequencing of DNA utilizes two basic principles: (i) nucleic acids are polymerized in 5' to 3' direction dependent on a template (use of the DNA polymerase involved in bacterial repair DNA synthesis) and (ii) an extendible primer has to provide a free 3'-OH end, hence, dideoxyribonucleotides, which lack a free 3'-OH end, will terminate the polymerization reaction. If the reaction is done separately for each of the four nucleotides, this generates four groups of DNA fragments of different lengths. A comparison of all the lengths allows the determination of the nucleotide sequence (imagine them stacked up vertically like organ pipes).

Question 4

A given DNA sequence can code for more than one protein by the following process:

(a) presence of multiple start sites (preceded by ribosome binding sites) in frame.

(b) there may be overlapping open reading frames that are shifted by one or two base pairs.

(c) differential splicing, i.e. the diverse utilization of exons for mRNA assembly.

Question 5

RNA virus replication depends on the action of RNA replicases which lack proofreading function. This favors genetic drift. Further, the genomes of many RNA viruses contain multiple linear fragments. If two different viral genomes are in the same host cell, recombination will facilitate genetic shift. Hence, it appears that RNA virus replication is not designed to maintain the blueprint for an 'organismal species' but to maintain the infectiousness of the genetic element which is dependent on the host cell machinery for the generation of progeny.

Answers to problems

Problem 1

The mutation is most likely in a gene encoding a *trans*-acting factor that plays some role in the transcription of both homologous structural genes. The mutation renders this factor nonfunctional.

Problem 2

This kind of disclosure requires that the locus has been mapped and the aim will be to isolate the gene(s), determine the nucleotide sequence and confirm the mutations as base substitutions. There is a good chance that the substitutions alter recognition sequences for DNA modifying enzymes such as methylases and endonucleases. If this happens, RFLP experiments can be employed to identify the genome with the mutant alleles.

Answers to summary worksheet

The emphasis of Part 1 was to introduce the basic structure and function of nucleic acids, the universal carrier of inheritable information. Chapter 6 in particular introduced principles and methods that allow characterization of nucleic acids in terms of a combination of **physical** dissection (nucleic acid sequence) and **positional** information (localization of individual genes). Nucleic acid analysis is based upon two major principles: the specificity of **endonucleases** for particular recognition sites and the template- and primer-dependence of nucleic acid **synthesis**. While this apears trivial in retrospect, the application of these principles to analytical investigation of DNA represented a novel integration of many not universally accepted hypotheses. Both basic techniques produce nucleic acid fragments of different **lengths** which are usually separated by agarose **gel electrophoresis**. The negative charge of DNA is proportional to its **length** due to fact that the **nucleosides** are polymerized via phosphodiester bonds hence electrophoretic separation occurs strictly by fragment size. This allows **standardization**. The length of a DNA fragment can be determined by using known DNA fragments such as a restriction digest of lambda phage DNA selected for by the desired range of fragment length resolution and running sample and standard side by side on a gel. The fact that **endonuclease** action is sensitive to mutation in the restriction site allows variant analysis independent of the **phenotype**. A mutation that abolishes or creates a restriction site constitutes a genetic polymorphism that is called **restriction fragment length polymorphism** (RFLP). RFLP analysis of chromosomes creates a **linkage** map while determination of the **nucleotide sequence** establishes a precise physical map. Both techniques have contributed to our understanding that the primary structures of DNA (**nucleotide** sequence) and protein (**amino acid** sequence) are colinear in **prokaryotes** but not **eukaryotes**.

Chapter 7: Messenger RNA

Answers to multiple-choice questions

1 (d); 2 (a); 3 (b); 4 (b); 5 (a,b,c,d,e); 6 (b,c); 7 (b); 8 (a).

Answers to concept questions

Question 1
Prokaryotic and eukaryotic mRNAs usually differ in the number of translatable cistrons in that eukaryotic mRNAs are exclusively monocistronic. Furthermore, eukaryotic mRNAs are polyadenylated at the 3' end and have a 7-methyl-guanidyl-phosphate cap at the 5' end. The trailer regions of eukaryotic mRNAs often contain specific destabilizing elements.

Question 2
Eukaryotic gene expression is compartmentalized in that synthesis and maturation of the mRNA are confined to the nucleus. The translation, however, is a cytoplasmic process and the transcribed 'message' has to be transported to the ribosomes outside of the nucleus. Since this transport of the message can be interfered with experimentally in many ways and because the half life of eukaryotic mRNAs is greater than that of prokaryotic mRNAs, it was possible to isolate the 'messenger' first from eukaryotic cells.

Question 3
mRNA constitutes only a small proportion (3–5%) of the total cellular RNA for two main reasons: First, due to the large number of ribosomes and the need for a constant pool of tRNAs, mRNA synthesis is proportionally less than that of the other RNAs. Second, mRNAs have a very short (~ 2 to 15 min in prokaryotes) or medium (4–24 h in eukaryotes) half life due to spontaneous degradation facilitated by endo- and exonucleases.

Question 4

The poly(A) tail of eukaryotic mRNAs is not encoded by DNA. RNA polymerase usually extends transcription hundreds of nucleotides further downstream from the location where the poly(A) is added to the 3' end. This is accomplished by a complex of enzymes that recognize and cleave the RNA at the polyadenylation consensus site and add approximately 200 adenylates by polymerization.

Question 5

Both prokaryotic and eukaryotic mRNAs must contain untranslated ribonucleotides (the leader) upstream of the translational start codon, because this provides stability and binding opportunities for the ribosome.

Answers to summary worksheet

Prokaryotic and eukaryotic mRNAs differ fundamentally in structure (see Parts 3 and 6) and in how they function in the process of translation (Chapters 8 and 9). Nevertheless, the products of **mRNA** translation in prokaryotic and eukaryotic cells are identical and they exist as polymerized amino acid molecules, called **polypeptides**. Because of the physical separation of DNA from the site of protein synthesis in **eukaryotic** cells, the foremost property of mRNA is that of an intermediate template for ribosomal amino acid polymerization. This physical separation, however, does not exist in prokaryotic cells and mRNA as a direct product of **transcription** is accessible to ribosomes already in its nascent form. Consequently, the stabilities of prokaryotic and eukaryotic mRNAs have evolved to be quite different with half lives of **15 minutes** and **4 to 24 hours**, respectively. While both mRNAs have untranslated regions at the 5' (**leader**) and 3' (**trailer**) ends, prokaryotic mRNA has additional untranslated sequences due to its **polycistronic** nature. These **intercistronic** sequences contain the ribosome-binding **Shine–Dalgarno** sequences upstream of the translational start codons of individual **cistrons** (coding regions, ORFs). The information stored in the **coding regions** of mRNA is literally translated into polypeptide sequence by aminoacyl-tRNAs, a process that is facilitated by ribosomes. Each aminoacyl-tRNA interacts with three consecutive nucleotides (= one codon) by recognizing complementarity (Watson–Crick basepairs) to its **anticodon** which exclusively determines aminoacyl-tRNA specificity in translation. Therefore, responsibility for the proper conversion of information stored in the **primary** sequence of DNA into functionality of protein rests in the enzyme that charges a tRNA with the cognate amino acid, which is **aminoacyl-tRNA synthetase** (Chapter 9).

Chapter 8: Protein synthesis

Answers to multiple-choice questions

1 (b); 2 (b); 3 (b); 4 (c); 5 (e); 6 (b); 7 (a,b,e); 8 (b,d); 9 (b); 10 (a).

Answers to concept questions

Question 1

N-Formyl-methionyl tRNA (fMet-tRNA) is the prokaryotic initiator aminoacyl-tRNA that recognizes both the AUG and GUG codons as translational start codons. Upon complex formation with IF-2, it is capable of entering the partial P-site in the small subunit.

Question 2

The eukaryotic 80S ribosome has about twice the mass of the prokaryotic 70S ribosome which is the result of a larger number of ribosomal proteins and rRNAs as well as the larger size of the rRNAs. Both the eukaryotic small (40S) and large (60S) subunits are larger than their prokaryotic counterparts (30S and 50S). While rRNA is the predominant mass component in both, the RNA content in prokaryotic ribosomes is higher. The prokaryotic ribosome has an additional E-site for facilitated exit of deacylated tRNAs.

Question 3

Major differences in the initiation of translation in eukaryotes and prokaryotes result from differences in the nature of the mRNAs as well as the ability of the small subunit to bind to the mRNA upstream of the start codon. Prokaryotic mRNAs are less stable and polycistronic, hence the prokaryotic initiation complex (IF-3, 30S subunit, mRNA, IF-2, GTP, fMet-tRNA) assembles after IF-3-mediated direct binding of the small subunit at the ribosome-binding site (via the 3' end of 16S rRNA). In contrast, several initiation factors are required for the preparation of eukaryotic mRNA (eIF-4, 4A, 4B) before the eukaryotic subunit initiation complex (SIC; 40S subunit, eIF-2, GTP, Met-tRNA) can bind to the mRNA cap (facilitated by the eIF-4 and eIF-3 factors). Once bound, the SIC scans downstream the mRNA until it encounters the first AUG codon.

Question 4

Petidyl transferase activity is a function of the large subunit. It resides in proximity of the amino acid stems of peptidyl-tRNA and aminoacyl-tRNA in the ribosomal P-site and A-site, respectively. The catalytic activity that transfers and covalently bonds the polypeptidyl chain to the amino group of the aminoacyl-tRNA in the A-site is believed to be a property of the large rRNA; however, the presence of several large subunit proteins is necessary.

Question 5

Termination occurs because there are no anticodons on aminoacyl-tRNA that are complementary to the three stop (termination, nonsense) codons. Hence termination *does not* involve tRNA. There is no amino acid that could be linked to the petidyl-tRNA in the P-site. One release factor (eRF), a protein, facilitates the termination, possibly by inducing translocation without preceding transpeptidation and deacylation of the C-terminal amino acid from the tRNA. The nascent polypeptide leaves the eukaryotic ribosome directly from the P-site into the cytoplasm (cytoplasmic ribosomes) or into the translocase tunnel (ER-ribosomes).

Question 6

The eukaryotic ribosome consists of the 40S (or small) subunit and the 60S (or large) subunit. The 40S subunit contains the 18S rRNA molecule and many ribosomal proteins. The 60S subunit contains the 5S, 5.8S and 28S rRNAs and many ribosomal proteins. The 5S, 18S and 28S rRNAs are transcribed from rDNA clusters in the nucleolus, and are separated by RNA-cutting endonuclease and exonuclease activities. The 5.8S rRNA is transcribed by a different enzyme in the nucleus. The ribosomal proteins are products of the expression of the respective nuclear DNA (transcription and processing of primary transcripts in nucleus, transport of the mRNAs into the cytoplasm, translation at 'free' ribosomes) that have to be transported through nuclear pores back into the nucleus where they associate with the rRNAs to form the 40S (18S rRNA) and 60S (5S, 5.8S and 28S rRNA) subunits. The assembled subunits are then transported into the cytoplasm.

Answers to problems

Problem 1

Because the gene has been sequenced and analyzed, the pre-mRNA structure is known. It is important to know the functional location of the protein to avoid problems with post-translational peptide processing. In the case of interrupted genes, cDNA has to be produced from the message which can be identified by *in situ* hybridization with exon based DNA probes. After cloning of the *Arabidopsis* gDNA/cDNA into yeast artificial chromosomes (YACs) as a tag/marker fusion for localization/selection, the gene could be expressed in yeast as a cytoplasmic or a secreted protein.

Problem 2

Inhibition of translation can be achieved by interfering with the individual steps in the process of ribosome function. This involves among others:

(i)   reduction of initiation factor availability—preventing ternary complex formation;

(ii) reduction of mRNA availability by antisense RNA causing RNA duplex formation;

(iii) selective blockage of P-site and/or A-site in the 30S subunit;

214

(iv)   introduction of missense suppressors;
(v)   preventing assembly of 30S and 50S subunits;
(vi)   preventing transpeptidation;
(vii) preventing translocation of the ribosome;
(viii) preventing ribosome dissociation.

Answers to summary worksheet

Protein synthesis is predominantly accomplished by the translation of **mRNA** into amino acid sequence by the **ribosome**. The ribosome is assembled from two subunits which are complexes consisting of a large highly conserved **ribonucleic acid** and various specific **proteins**. Ribosomal translation of information derived from the DNA by transcription into polypeptide sequence occurs similarly in both prokaryotic and eukaryotic cells: the paradigm of ribosomal protein synthesis lies in the guided base-pairing of mRNA ribonucleotide **triplets**, or codons, with complementary **anticodons** of aminoacyl-tRNAs. Thereby the amino acid primary sequence is established following the mRNA codon sequence in 5' to 3' direction. Crucial steps in this process are the finding of the first codon of the information-bearing stretch on the mRNA (the **open reading frame** or cistron), the supply with needed aminoacyl-tRNAs, the formation of the peptide bonds between the **amino acids** and the finding of the stop signal which ends the amino acid polymerization process. Ribosome assembly always begins with the facilitated binding of the small subunit to the **mRNA** followed by the binding of the initiator tRNA to the partial **P-site** before small and large subunits form the functional ribosome. All these steps occur in a distinctly different manner in prokaryotic and eukaryotic cells. While the eukaryotic facilitator of cytoplasmic translation, the **80S** ribosome, cannot read the start position directly and has to scan along the mRNA starting at the cap at the **5' end**, the prokaryotic **70S** ribosome reads the translational start position directly. This difference already obviates the participation of different **initiation factors** in eukaryotes and prokaryotes. Eukaryotic mRNA scanning requires binding of factors eIF-4, 4A and 4B and the absence of extensive **secondary** structures. The direct 'reading' of the translational start codon by the prokaryotic ribosome is also necessary because of the **polycistronic** nature of prokaryotic transcripts: every intercistronic (intergenic) sequence must contain a **ribosome binding site** in order to attract the ribosome for assembly upstream of the individual cistrons. Specific initiation factors bind to the small subunits to prevent aggregation of the two subunits without binding to translatable message. Other factors are employed to stabilize the functional ribosome once it has assembled and translation has successfully initiated. Study of the ribosomal **transpeptidation** activity has contributed to a completely new chapter of biology with a huge evolutionary dimension: the catalytic activity is suspected to not be of enzymic nature, instead, it might reside in the rRNA. This discovery together with that of autocatalytic pre-mRNA processing (splicing, see Parts 5 and 6) implies that our perceived picture of a cellular world may be just one form of how life is organized at the molecular level.

Chapter 9: Interpreting the genetic code

Answers to multiple-choice questions

1 (a); 2 (c,d); 3 (a); 4 (c,d); 5 (a,b); 6 (b); 7 (c); 8 (b,c,d,e); 9 (b,c,d); 10 (a).

Answers to concept questions

Question 1

Two classes of ten aminoacyl-tRNA synthetases each make functional aminoacyl-tRNA. In a two-step process, a particular amino acid is first phosphorylated to aminoacyl-AMP and then the cognate tRNA is charged with the amino acid (breakage of the phosphoester bond). The accuracy of this reaction is the foundation for the translatability of mRNA following the genetic code.

Question 2

The wobble hypothesis summarizes what is also known as third base pair degeneracy. The reason for this is that the first base in the tRNA anticodon is flexible due to the loop structure and allows stable pairing with either all four bases or the two purines or pyrimidines. Look at the color coding in Table 9.3.

Question 3

Ile-tRNA recognizes AUU, AUC and AUA but discriminates against AUG. To allow the 'odd' pairing of Ile-tRNA, the first anticodon base is inosine (instead of adenine) which can pair with U, C and A but not G. Look at Figure 9.4.

Question 4

The presence of strong missense suppressor is a big problem for the cell. While it helps to suppress a particular mutation and rescue a small number of proteins, it will probably cause amino acid substitutions in many wild-type proteins. Hence, a repressor simultaneously functions as a mutator.

Question 5

Kinetic proofreading refers to the fact that aminoacyl-tRNA synthetases have specific affinity for their cognate tRNAs. Noncognate tRNAs bind very slowly to the enzyme's binding site and, consequently, cannot cause the conformational change (necessary to facilitate the bonding between tRNA and aminoacyl-AMP) quickly enough before the tRNA dissociates uncharged.

Answers to problems

Problem 1

If suited for the organism, nonsense suppressor tRNAs could be introduced/generated which will cause readthrough on the native transcript. Alternatively, a recombinant translational fusion construct could be generated and expressed from a plasmid (e.g., in bacteria) or incorporated into the genome by homologous recombination. The latter strategy can be also used for eukaryotes such as diploid yeast which allows functional analysis after haploid segregation.

Problem 2

Mutational drift follows a bias to either higher A+T or higher G+C content of the third position of the codons in AT-rich or GC-rich organisms, respectively. Recombination of acquired DNA seems to follow a similar bias. Codon usage across the genome is usually correlated with the abundance of different tRNA species and shaped by the selection for particular codons that are more efficiently and/or accurately translated.

Answers to summary worksheet

The genetic code is universal and encrypts everything necessary to maintain cellular life. The organization of the four **ribonucleotides** in groups of three, constituting a codon, gives 64 possible combinations which is considerably more variability than necessary for encoding the 20 **amino acids**. This has the consequence that almost all amino acids are represented by more than one codon in the range from two to six. Interestingly, this is mostly accomplished by variation in the **third** base of the triplet by a lack of discrimination among the pyrimidines (U, C) or purines (A, G). In fact, because methionine and tryptophan are the only amino acids with a unique codon which ends with a G in the third position, C, U and A in the third position never define a particular amino acid. This third base pair **degeneracy** is enabled by the G•U base pairing between the **third** codon base and the **first** anticodon base, also called 'wobbling' in that 'synonym' triplets code for the same amino acid. While the start codon of an open reading frame is defined and represented by AUG in eukaryotes and AUG and GUG in prokaryotes, the **stop signal** is not translated via complementary base-pairing. Instead, three of the 64 triplets, UAG (amber), UAA (ochre) and UGA, regularly do not match any tRNA anticodon, hence there will only be nonspecific competition for tRNA entrance into the ribosomal **A-site**. This competition is regulated by so-called termination factors, proteins that prevent nonspecific binding and facilitate **deacylation** of the peptidyl-tRNA in the **P-site**. While the transpeptidation could be facilitated by rRNA (Chapter 8), the master players in the information game are enzymes that understand both the amino acid and the ribonucleotide languages, the **aminoacyl-tRNA synthetases**. Two families of this enzyme with ten members each control the acylation of tRNA with the correct amino acid. Because this process requires the highest accuracy and fidelity besides that of DNA replication (Part 4), efficient mechanisms of **kinetic** and **chemical proofreading** have evolved. As potential vehicles of evolution, natural **suppressors** (modified anticodons in length or sequence) help to compensate for missense or nonsense mutations. In addition, they provide an efficient tool for differential translational regulation by providing **frameshift** suppression.

Chapter 10: Protein localization

Answers to multiple-choice questions

1 (c); 2 (b); 3 (b,c,e); 4 (c); 5 (a); 6 (b); 7 (b); 8 (b,c); 9 (c,e); 10 (c,d).

Answers to concept questions

Question 1

Unprocessed polypeptide chains in both eukaryotes and prokaryotes contain at their N-terminus at least a methionine which is usually cleaved by aminopeptidase. Polypeptides that reside in the organelles or the nucleus or that will be incorporated or transported across the plasma membrane contain N-terminal signal sequences. These are recognized by the translocation machinery and usually cleaved after passage of the membrane. The amino acids are usually hydrophobic or, at least, unpolar.

Question 2

ER proteins have a C-terminal recognition sequence, KDEL, which is recognized by a specific receptor/transport protein in the secretory system. Upon encounter, the receptor–protein complex travels back to the ER.

Question 3

Chaperones of the Hsp70 family fulfil the following tasks in healthy and stressed cells:
(a) stabilize newly synthesized proteins;
(b) preserve folding competence of polypeptides;
(c) preserve translocation competence of nascent polypeptides;
(d) facilitate protein degradation;
(e) promotes protein oligomer assembly/disassembly;
(f) reactivate thermally inactivated proteins, etc.

Question 4

Co-translational translocation describes the process in which the nascent polypeptide chain is translocated across the membrane as it is emerging from the ribosome. This is typical for RER ribosome polypeptide synthesis. Post-translational translocation is the process in which the

218

polypeptide is synthesized and either folded and unfolded (by chaperones) or kept in a translocation competent state (by chaperones). Translocation occurs later independent of the synthesis.

Question 5

Look at Figure 10.31; the experiment was aimed at finding evidence for the presence of more than one nuclear translocation factor. Secondly, the experiment addressed the problem of compartmentalization and proved that various cytosolic factors including ATP are necessary for docking and translocation of proteins into the nucleus.

Answers to problems

Problem 1

The mRNA coding for polypeptide A is produced in the nucleus while the mRNA encoding polypeptide B is a product of mitochondrial transcription. The primary transcript A is pre-mRNA which needs to be polyadenylated and capped first and then processed to mRNA by splicing. Translation of ORFA is facilitated by 'free', cytoskeleton-bound ribosomes in the cytoplasm and polypeptide A must contain two consecutive N-terminal signal sequences necessary for targeting to the mitochondrial matrix (first) and intermembrane space (second). A mutation creating the C-terminal sequence 'KDEL' should have no influence on polypeptide A protein trafficking because redirection of polypeptides bearing KDEL termini to the ER lumen requires that the polypeptides reside in the secretory system which they enter by co-translational translocation. Instead the nascent polypeptide is chaperoned on its path to the mitochondrial membrane, after translocation within the matrix and the intermembrane space (thus, cells defective in heatshock protein biosynthesis also have defective organellar protein uptake). ORFB is translated from the respective mRNA by mitochondrial ribosomes in the matrix, chaperoned and translocated to the intermembrane space, where the two polypeptides assemble.

Problem 2

Only sequences B and D are suitable as signal peptides because they have a positive N-terminus followed by a stretch of 21 largely hydrophobic amino acids with a potential aminopeptidase recognition site. In contrast, sequences A and C lack the hydrophobic stretch of amino acids, hence they will probably not be recognized by SRP or the SecA protein.

```
(A) MAADSQTPW-LLTFESLKCLHWPQEDQRAGA-LFGRAKLTR   not suited
(B) MMITLRK-LMPLAVAVAAGVMSVFAQAMA-VETDRAKLE     suited
(C) MPLLNWSRH-MVDDTAAKLITVPTRHWLVYA-TDTLTRDNG   not suited
(D) MPEGRRLRR-ALAIALVPLALVAVTGLLMMA-KEQQMGNEI   suited
```

219

Answers to summary worksheet

Proteins secreted by prokaryotes include a wide range of toxins, degradative enzymes and other factors contributing to stress tolerance, virulence and pathogenicity. This is accomplished by three major, functionally independent **secretion** pathways, called type I, II or III secretion systems. While many proteins are exported across the plasma membrane via the general secretory pathway (Sec pathway, type II), nonproteinaceous compounds, extracellular proteins from Gram-negative bacteria and proteins lacking an N-terminal signal sequence cannot use the Sec pathway. Instead, this is accomplished by two additional major groups of dedicated export systems: the ATP-binding cassette (ABC) transporters (type I) and the type III transporters. ABC transporters operate in both prokaryotes (e.g., histidine uptake and hemolysin export) and eukaryotes (e.g., the P-glycoprotein multidrug resistance pump). Type III secretion appears to occur only in prokaryotes and utilizes components related to the synthesis machinery for type IV pili and flagella to export pathogenicity determinants (e.g., 'harpins', Yops) or enzymes (e.g., pullulanase). This text focuses on the signal sequence-**dependent** type II general secretory pathway in prokaryotes (SecABDEFY, signal peptidase) as well as the post-translational (into organelles) and co-translational (into the RER) export of **eukaryotic** proteins. Nascent polypeptides destined to leave the compartment in which they were synthesized need to be kept in a **translocation-competent** conformation which is secured by molecular chaperones of the **Hsp70** family. Similarly, molecular chaperones of the **Hsp60** family facilitate proper protein folding after translocation. The N-terminal eukaryotic signal peptide is recognized by a **signal ribonucleoprotein** (SRP) which facilitates contact between the polypeptide and the translocase via specific interaction with a membrane receptor ('docking') protein. In prokaryotes, the roles of SRP and docking protein are combined in the form of the **SecA** membrane protein. 'Delivery' of a translocation-ready polypeptide costs the cell a great deal of energy: GTP hydrolysis is required for dissociation of **SRP** and insertion of the signal sequence into the translocase and ATP hydrolysis is necessary for release of the immature peptide from the **SecA** protein. In addition, protein unfolding as well as initial prevention of polypeptide folding by **molecular chaperones** requires additional hydrolysis of ATP. Proteins destined not for export but for incorporation into membranes, initially use the signal recognition/translocation mechanism but terminate translocation by so-called **stop-anchor** sequences which remain in the membrane as 21 amino acid-long **membrane spanning domains** (MSDs). Dependent on the number of MSDs in a polypeptide, they will incorporate into the membrane following the 'sewing machine principle', thereby creating intracellular and extracellular loop domains of transmembrane proteins. Least understood, so far, is the protein export from the cytoplasm into the nucleus (and vive versa), which is assumed to proceed through **nuclear pores**. These pores represent molecular mass sieves that do not discriminate against passage of molecules with masses smaller than 50 kDa. Larger nuclear proteins require gated **active transport** which consists of a **docking** phase and an ATP hydrolysis-dependent **translocation** phase. In addition, nucleoplasmic Ran-GTPase activity is required. Besides translocation, recycling of various overproduced, misdirected or damaged proteins follows protein biosynthetic activities. This is accomplished by **molecular chaperones** as well as **ubiquitin** which marks these proteins for degradation and directs them to the proteolytically active **proteasome**.

Chapter 11: Transcription

Answers to multiple-choice questions

1 (c); 2 (e); 3 (a); 4 (a); 5 (c); 6 (c); 7 (b); 8 (c); 9 (a); 10 (b).

Answers to concept questions

Question 1

Transcription is an enzyme-facilitated process. Substrates are the four nucleoside triphosphates (NTPs) and DNA serves as a template. DNA-dependent RNA polymerases (one in prokaryotes, three in eukaryotes) catalyze the polymerization of the NTPs thereby creating ribonucleic acids as products.

Question 2

Bacterial DNA-dependent RNA polymerase core enzyme consists of four subunits ($2\times \alpha$, β, β') which is to be complemented by the sigma factor (σ subunit) to create the holoenzyme. While core enzyme can catalyze NTP polymerization, only holoenzyme can initiate transcription. The key step is the specificity of sigma-factor to specific promoter consensus sequences upstream of the transcriptional startpoint. This allows tight binding of holoenzyme thereby creating the closed binary complex. The critical interaction is thus the interaction of RNA polymerase with the DNA. This is completely different in eukaryotic cells, where TBP containing transcription factors interact with the DNA, allowing other factors to bind and build the initiation complex. This complex recruits RNA polymerase, hence the critical interaction is the interaction of RNA polymerase with proteins.

Question 3

As a result of template and coding strand separation during transcription, single stranded DNA is protected in that the polymerase covers the entire transcription bubble: from the unwinding point to the rewinding point. In contrast to replication, participation of single strand binding proteins is not necessary.

Question 4

Sigma factors (except for RpoN) have domains that recognize DNA at the promoters, however, as free proteins sigma factors are in a non binding conformation. When sigma factor binds to the core enzyme, its N-terminus frees the DNA-binding domain in a conformational change of the protein. This regulated access of sigma factor to the promoter prevents free sigma factor from binding at the promoter sites thereby blocking initiation of transcription by a functional holoenzyme. In addition, it prevents that the concentration of available sigma factors for holoenzyme formation does not become too dilute: each cell contains approximately one sigma factor per three core polymerase enzymes.

Question 5

Mutations in regulatory sequences such as the *cis*-acting promoter, do not necessarily prevent transcription of the respective transcriptional unit. However, the initiation of transcription may become less efficient and the transcription of the adjacent genes becomes reduced. Such a mutation is called a down mutation. If mutations change the promoter sequence so that it initiates transcription more efficiently, they are called up mutations.

Question 6

Gyrase is a topoisomerase (type II) that has the ability to introduce negative supercoils into DNA. This will increase the supercoiling of the DNA. Topoisomerase I relaxes highly negatively supercoiled DNA. Hence, a single topoisomerase mutant will result in an irreversibly changed DNA topology which has dramatic effects on transcription. Because of the functional topoisomerase I, gyrase mutants will eventually have a too much relaxed DNA, that will prevent successful transcription. Double mutants, if not exposed to DNA-damaging environmental stress (that may cause DNA single strand breakage) will maintain the supercoiling of the DNA and can survive.

Question 7

The selectivity of sigma factor for specific promoter consensus sequences allows for differential gene expression. This is accomplished in that association of core enzyme with different sigma-factors creates different holoenzymes. If the synthesis of some sigma-factors is not constitutive but inducible dependent on environmental or developmental signals, differential gene expression is possible. A good example is the regulation of sporulation in endospore-forming bacteria.

Question 8

While bacterial sigma factors such as the *Escherichia coli* sigma factor σ^{70} (RpoD) bind DNA as part of the polymerase holoenzyme, the *rpoN* gene-encoded sigma factor, σ^{54}, has DNA-binding properties enabling it to bind DNA independent of core enzyme. Hence it appears that the RpoN protein functions more like a eukaryotic transcription factor, recruiting core enzyme to the promoter in a protein–protein interaction.

Question 9

Core enzyme has a high intrinsic affinity for DNA and it is currently assumed that only a very small amount of RNA polymerase is free. Whether before or after the initiation of transcription, RNA polymerase appears to be randomly distributed across the genome bound to so-called loose binding sites or in the process of RNA synthesis. If core enzyme is associated with sigma-factor, holoenzyme has high affinity to tight binding sites only, which allows initiation of transcription (formation of the open binary complex). Thus, the equilibrium between loose and tight binding of core enzyme to DNA reflects the equilibrium between core enzyme and holoenzyme both of which are bound to DNA. This makes transcriptional initiation much more efficient and allows the regulation by operon repression/derepression as well as differential transcription based upon sigma factor specificity.

Question 10

Review termination of transcription in prokaryotes.

(a) Two basic mechanisms account for the termination of transcription in prokaryotes: (i) termination by intrinsic terminators at certain sites without the facilitation by any other factor and (ii) Rho-factor (protein) dependent termination.

(b) Translation can regulate transcriptional termination by attenuation which occurs for instance in the expression of amino acid biosynthesis genes. Translation of the leader peptide regulates transcription of structural genes downstream. Important in the biosynthesis of amino acids (e.g., tryptophan). Transcription and translation in prokaryotes occur virtually simultaneously, hence the translating ribosome is chasing the transcribing DNA-dependent RNA polymerase. If the ribosome is furnished with all necessary aminoacyl-tRNAs, it will translate the leader into a leader peptide and terminate translation at the stop codon of the leader ORF. Here *attenuation* occurs (transcription of the structural genes downstream stops) because the newly formed mRNA is free to form a terminating 3:4 stem–loop (complementary base pairing), which 'bumps off' the DNA-dependent RNA polymerase. If any amino acid is in short supply thereby causing a shortage of the respective aminoacyl-tRNA, the ribosome pauses before the stop codon at the codon in the leader ORF that codes for the amino acid in short supply. Here an alternative 2:3 stem–loop can form, the *antiterminator*. This stem–loop is not a terminator signal for polymerase and it prevents the terminating 3:4 stem–loop from forming thereby allowing transcription of the structural genes to proceed.

(c) In prokaryotes, the processes of transcription and translation occur virtually simultaneously, meaning the ribosome is chasing the DNA-dependent RNA polymerase. The Rho-factor has an RNA-dependent ATP(hydrol)ase activity which acts on transcription by separating the RNA–DNA hybrid in the transcription bubble. To do this it has to move along the transcript $(5' \rightarrow 3')$ to catch up with DNA-dependent

223

RNA polymerase. If ribosomes are in its way, termination by the Rho-factor is not going to occur.

(d) A very frequently employed mechanism is the process of *antitermination* (see phage genetics). Antiterminators are proteins that recognize antitermination sequences in the DNA upstream of the termination sequence, the antiterminator utilization sites (e.g., the *nut* sites in lambda phage DNA). At such a sequence, the antiterminator facilitates the interaction between antiterminator utilization substances (e.g., the Nus-proteins in lambda phage gene expression) and DNA-dependent RNA polymerase thereby causing a readthrough beyond (downstream) the terminator sequence.

Answers to problems

Problem 1

```
GAATTCCGGCGGCGGGGTTTTCATTATTATGATCGTTGACATGGGACAAGGGGCTCCTATAATAGGTGC
EcoRI             -43              -35                      -10
ATTTGTAGGGGATGTGGCCACATGATAGCGCGTCGAGGCATTCAGGACCGATATTGTCATATTGCCAGC
+1    S/D              M  I  A  R  R  G  I  Q  D  R  Y  C  H  I  A  S
ATGCTGCAGCGCGCCAACTTAGTTACAACTTATACGATGATTTTAAAAAAAGAATATCAAATAGCCATC
M  L  Q  R  A  N  L  V  T  T  Y  T  M  I  L  K  K  E  Y  Q  I  A  I
CACCCGAAAGACCACCGCAAGTATCGGTACGATCGTACCAGCACACACCACAGTCGCCAGTCTGTTACC
H  P  K  D  H  R  K  Y  R  Y  D  R  T  S  T  H  H  S  R  Q  S  V  T
AAATAATAAAGTTGATAATTAATTCACATCTACCTGGGTAACCCAGGTAGATGTGAATTTTTAGAATTC
K  *  *                    ------------------> <------------------   EcoRI
```

Problem 2

It is of utmost importance to faithfully maintain the genetic information at the DNA level and thus DNA-dependent DNA polymerases have proofreading function. The products of transcription and RNA replication, RNAs, are usually short-lived and transcriptional mistakes are 'repaired' by the natural degradation activity of RNAases. Nonfunctional products are replaced by correct copies. From an evolutionary perspective, RNA viruses seem to have the advantage of antigenic drift for evasion of host defense mechanisms such as the immune system of vertebrates ('employed' by HIV and influenza virus) or nucleic acid restriction systems.

Problem 3

Prokaryotic gene expression is a complex and highly integrated process in space and time. The directedness of transcription and translation is important in order to maintain the reactions collision-free. Replication of circular DNA, which occurs simultaneously on both strands in the 5' to 3' direction is slowed and halted by specific stop sequences. During translation, the ribosome always chases RNA polymerase. Because mRNA is single-stranded and subject to rapid degradation, it would make little sense to 'allow' translation to occur in the 3' to 5' direction.

Answers to summary worksheet

Prokaryotic DNA-dependent RNA polymerase interacts directly with DNA. One of its subunits, the **sigma factor**, recognizes and binds to a highly conserved, *cis*-**acting** element: the **promoter** (of transcription). **Core enzyme**, made up of four subunits ($2\times \alpha$, $1\times \beta$ and $1\times \beta'$), and sigma (σ) factor assemble at the promoter to the **holoenzyme**, Eσ. Both β and β' subunits constitute the catalytic center and, not surprisingly, share a high degree of similarity with the large subunits of eukaryotic RNA polymerases (see Chapter 28). The α subunit dimer appears to be involved in holoenzyme assembly as well as in promoter recognition. RNA polymerase recognizes promoters with different **consensus sequences** due to the specificity of different **sigma factors**. These consensus sequences are two conserved blocks of 5 to 8 nucleotides located approximately 10 and 35 nucleotides upstream of the mRNA start site, the **transcriptional startpoint** (*tsp*) located at '+1' in the DNA. The promoter consensus regions are, therefore, called '**−10**' and '**−35**' regions. While the '**−35**' **region** is implicated primarily in recognition and binding of RNA polymerase, the '**−10**' **region**, also known as **Pribnow-Box**, is thought to be the location where the DNA duplex is melted. Once bound, holoenzyme and promoter form a closed binary complex which changes into an **open binary complex** upon melting of the duplex. This 'tight binding' is usually irreversible. Mutagenesis of the promoter region will alter the binding of **RNA polymerase** to **DNA** and, consequently, initiation efficiency. Dependent on whether transcription is enhanced or reduced, these mutations are called **up** or **down** mutations.

RNA polymerase will facilitate the placement of ribonucleotides at the **transcriptional startpoint** which are complementary to the template DNA. The formation of phosphodiester bonds between the ribonucleotides creates the **DNA–RNA hybrid** which together with RNA polymerase is also called **ternary complex**. The nascent RNA can grow up to **9 nucleotides** long before the transcription complex needs to move forward (5' to 3') on the DNA. If the chain is not released prematurely (called **abortive initiation**), the chain will grow as the holoenzyme moves along the **template strand**; transcription has been successfully initiated. At this stage, the **elongation complex** has been created and **sigma factor** will be released.

Elongation is accomplished by **core enzyme RNA polymerase**. RNA polymerase moves along the DNA reading the **template strand** of DNA in **3' to 5'** direction thereby polymerizing a nascent RNA in the **5' to 3'** direction which is identical to the **coding strand** of DNA. Because DNA exists conformationally in a double-helical form, the DNA has to be unwound and rewound which creates the so-called **transcription bubble**. This is accomplished by **RNA polymerase** at the **unwinding** and **rewinding** points. During elongation, the melted single-stranded DNA in the transcription bubble is protected by **RNA polymerase**. In contrast, the

225

nascent RNA emerges in a **5' to 3'** direction from **RNA polymerase** as a **free single strand**. Elongation will occur as long as RNA polymerase does not encounter a **termination** signal. This signal can be either an **intrinsic terminator stem–loop** structure or the **Rho** protein. The first occurs in the nascent RNA and stalls the enzyme. If the DNA template at this (growing) point contains a stretch of **adenylates**, the **RNA–DNA hybrid** interaction is extremely weak and both **nucleic acids** dissociate. This leaves the RNA polymerase without a free **3' end** for further extension the transcription is terminated. Transcription and translation are tightly coupled in prokaryotes, hence a normal situation would be that a translating **ribosome** chases a transcribing **RNA polymerase**. Another way of transcriptional termination is accomplished by the **Rho protein** that has to bind to the **5' end** of the **primary transcript** and sneak up in order to interact with RNA polymerase. Because of the tight coupling of transcription and translation **Rho**-dependent termination of prokaryotic transcription is rare.

Chapter 12: The operon

Answers to multiple-choice questions

1 (d); 2 (a); 3 (b); 4 (a); 5 (e); 6 (a); 7 (d); 8 (a); 9 (b); 10 (a,d).

Answers to concept questions

Question 1
Both X-gal (5-bromo-4-chloro-3-indolyl-β-D-galactoside) and IPTG (isopropyl-β-D-thiogalactoside) have the galactoside moiety of lactose acts as inducer in β-galactosidase expression (by derepression). The major difference between the two compounds is that IPTG cannot be utilized as a substrate by the enzyme, hence it remains available as inducer. X-gal is utilized and the cleaved chromophore contributes to the blue color indicating a functional β-galactosidase (in 'blue–white screening').

Question 2
A eukaryotic merodiploid is a cell that contains a complete set of chromosomes and one additional single homologous chromosome. A merodiploid prokaryote which has usually only one chromosome, contains an additional copy of at least one homologous transcriptional unit. In this case two mutants were complemented in *trans* by complete transcriptional units, hence they were merodiploids. One mutant (A) was successfully complemented (reverted to the wild-type) while the other (B) was not. The mutation in A is most likely in the *lac* repressor gene, *lacI*. However, since the Lac repressor is a tetramer, the mutation was most likely not in the region responsible for DNA-binding. This is because even though expression of the wild-type copy makes functional LacI, the presence of

'faulty' copies will prevent binding of any repressor. The mutation in B was most likely in the operator disallowing the LacI repressor to bind. Since operator mutations are *cis*-acting, the merodiploid Lac phenotype remains constitutive.

Question 3

One of the rare cases of positive control in prokaryotes is the catabolite repression found in *Escherichia coli*. This system enforces preference for the utilization of the catabolic substrate glucose over others that are frequently present in the bacterium's microenvironment such as lactose, galactose and arabinose (glucose catabolism is bioenergetically more economical). The mechanism depends on the fact that high glucose levels reduce the activity of adenylate cyclase, the producer of cyclic AMP. Low cAMP levels are insufficient to bind and activate the 'catabolite activator protein', CAP. The cAMP–CAP complex binds in the vicinity of promoters where CAP can interact directly with RNA polymerase (e.g., *gal* and *lac*). In rare other cases such as *ara*, binding occurs further upstream; however, CAP and polymerase appear to bind to the DNA on the same side and the CAP-induced bend may allow direct interaction. CAP–polymerase interaction represents a positive control because at low glucose levels expression of other catabolite operons is induced.

Question 4

The regulon described in the book is the *trp* regulon (Fig. 12.19). A regulon is usually named after the repressor gene whose product controls multiple promoters by repression. In the case of *trp*, the *trp* repressor, TrpR, controls
(a) the operon with the structural genes for tryptophan biosynthesis;
(b) expression of one of the genes coding for one enzyme in aromatic amino acid synthesis;
(c) the expression of the *trpR* gene, hence exerts autogenous control as found for oxyR or the lambda repressor;
A mutation in *trpR* has thus direct effects (constitutive phenotype) and is not pleiotropic like a modulon mutation could be.

Answers to problems

Problem 1

Create EPS mutants and do complementation analysis in *trans*. Alternatively, use the sequence of one of the genes for the engineering of a probe. Isolate total RNA for Northern analysis.

Problem 2(a)

If the expression host is able to grow in the presence of a high tryptophan concentration, construct a transcriptional fusion to a *trp* promoter plus leader peptide and attenuator element and grow at high tryptophan concentration. For expression, resuspend in low tryptophan medium thereby achieving control by attenuation.

If the host does not grow well in the presence of a high tryptophan concentration, fuse the DNA to a lambda phage P_L promoter and express lambda repressor (*cI*) under control of the *trp* promoter. The latter is repressed by *trp* repressor in the presence of its co-repressor, tryptophan.

Problem 2(b)

Construct antisense RNA encoding DNA and place under control of the same promoter upstream of DNA.

Problem 2(c)

Find a unique restriction site within the gene and outside of the gene. Cut and ligate both pieces in opposite direction. Plasmid-amplify to accumulate desired amounts of DNA. Cut and religate.

Problem 3

OxyR must be under positive autogenous control. DNA binding conformation is controlled by the oxidation state of the protein.

Problem 4(a)

Autophosphorylation is induced upon ligand (nutrient)–receptor interaction. Hpk phosphorylates response regulator (CheY) which becomes able to interact with flagellum proteins. The effector reaction is a protein–protein interaction.

Problem 4(b)

Autophosphorylation induced upon ligand(toxin)–receptor interaction. Hpk phosphorylates response regulator which becomes able to bind to operator elements which in turn initiates transcription of toxin resistance genes. The effector reaction is RNA-polymerase recruitment.

Problem 4(c)

Autophosphorylation is induced upon ligand(nitrogenous compounds)–receptor interaction. Hpk phosphorylates response regulator, NtrC, which becomes able to bind to enhancer element. Enhancer-bound NtrC interacts with σ^{54} (*rpoN*) in the closed binary complex thereby triggering a shift to a transcription-ready open state. The effector reaction is RNA-polymerase isomerization.

Problem 5

What is needed is a medium strength operon promoter yielding transcript that contains all ORFs. The intergenic region between ORFs must contain another promoter that drives transcription of the catalytic site-containing subunit gene. Therefore, the latter must not be the first gene in the operon.

Problem 6

Use DNA footprinting (cutting free DNA with DNase I) of the region of at least -100 to $+50$. You must compare at least: free DNA (and a non-binding protein like BSA); DNA with cell extract from *lacI⁻* mutant (RNA polymerase binds); DNA with fluid from wild-type (lac repressor binds) and a marker such as a Maxam–Gilbert G+A sequencing reaction. If similar fluids are obtained from glucose-starved, lactose-grown cultures, a cAMP–CAP footprint overlapping with the RNA-polymerase footprint becomes evident.

Answers to summary worksheet

The control of gene expression in single-celled prokaryotes serves mainly to empower the cell with quick and flexible responses to environmental signals and cues. This is in contrast to multicellular eukaryotes in which the regulation of gene expression follows genetic programs in processes of differentiation and development. Prokaryotic gene expression responds to abruptly fluctuating physical and chemical parameters including excess or gradual depletion of essential nutrients (nutrient stress or starvation), exposure to noxious chemicals (chemical stress) as well as competition for surface sites (physical stress). Against intuitive expectations, well nurtured, rapidly growing prokaryotic cells are highly susceptible to environmental stress, while starved cells have a high stress tolerance. Remember that stationary phase, starved cells are physiologically 'sedate', maintain only one copy of their genome and a reduced rRNA content with the consequence of an overall low rate of protein synthesis. Naturally, this provides fewer opportunities for interference of stresses with metabolic and gene expression activities. The magic that prokaryotic transcriptional and translational regulation can work becomes evident in that stressed or resting bacteria can switch to high levels of gene expression and metabolic activities within just minutes; even entire populations are able to respond synchronously (e.g., motility waves of swarming bacteria such as *Proteus* or *Myxococcus*). These well-defined and specific responses to environmental stresses can be explained only by the existence and operation of multigene and global control systems that are organized along strict hierarchical principles.

The smallest expression unit is the **operon**, which represents a cluster of one or more contiguous genes that are transcribed coordinately from a single promoter. Transcription is terminated downstream from the last operon member gene by mechanisms such as intrinsic termination or the action of the **Rho** protein. The product, a **polycistronic** (multi-genic) mRNA, is usually translated into polypeptides that contribute to a common functional goal (i.e., a biochemical pathway; synthesis and secretion of a defense enzyme). Hence, which and how much of a particular mRNA is going to be produced is determined by the process of **transcription** initiation. This represents a crucial control point. This 'up-front' regulation approach is accomplished by processes of RNA polymerase recruitment and isomerization which involves the binding of *trans*-acting factors with *cis*-acting elements as well as the interaction between proteins bound to DNA rather far apart. These transcriptionally regulatory *cis*-

elements are the **promoter**, **operator** and **enhancer** sequences. The first is selected by the specificity of **sigma factor**, which allows initiation of transcription of selected operons based upon sigma-factor availability. The second provides an opportunity for DNA-binding proteins to bind and, consequently, positively or negatively affect RNA-polymerase binding to the promoter. These DNA-binding proteins are **repressors** and **activators** (which exert negative and positive control of transcription, respectively). Both can be involved in the induction and repression of transcription which is initiated by **inducers** and **co-repressors**, respectively. By interaction with an active repressor, an inducer weakens the interaction between repressor and **operator** and thus de-represses transcription. In contrast, a co-repressor enables stable binding interaction between repressor and operator; initiation of transcription is prevented. Similarly, activators can be activated and deactivated resulting in induction and repression, respectively.

More complex and hierarchically higher regulatory units can be defined along genotypic or phenotypic lines. Genotypically, if multiple, unlinked operons are controlled by the same protein repressor, they constitute a **regulon**. Member operons usually contribute to a more complex (but single) metabolic pathway or function and the **regulon** is named after the gene encoding the repressor (e.g., OxyR regulon). In a **modulon**, member operons are controlled by a pleiotropic regulatory protein and contribute to multiple pathways (e.g., cAMP–CAP modulon). In addition, member operons may still be governed by individual regulatory proteins. Phenotypically, a set of independently co-induced (derepressed or activated) or co-repressed operons that respond to a single environmental signal constitute a **stimulon**. Stimulon member operons may belong to various regulons and modulons, and thus do not share a regulatory protein (e.g., hydrogen peroxide stimulon). This means that membership of an operon in different regulatory units creates overlap in the response where different stimuli/stresses cause related response phenotypes. This is particularly evident in the coordinated regulation in operon networks such as the multigene systems responsible for cell division or sporulation. Other global control systems such as the heat shock or SOS response are often called 'adaptive response systems' and they entail multi-pathway responses to a specific, well-defined stress.

Interestingly, transcriptional activity can also be regulated by the ribosome. This is caused by the close physical and temporal association of transcription and translation in prokaryotic cells. In a process called **attenuation**, an RNA-polymerase-chasing ribosome determines the timing of intrinsic termination by being close to core enzyme (antiterminator can not form, terminator formation yields termination) or by lagging behind (antitermination allows core enzyme to continue transcription). Slowing of the ribosome stems from insufficient availability of selected aminoacyl-tRNAs: it is logical that this mechanism regulates the expression of genes that encode enzymes involved in biosynthesis of the very amino acids that the cell needs to charge the respective tRNAs. The ribosome facilitates the response to amino acid starvation also posttranscriptionally. Association with the RelA protein near the A-site allows the synthesis of (p)ppGpp upon encounter of uncharged tRNAs. While the ribosome becomes idle, (p)ppGpp effects various changes in the cell's overall transcriptional activity that are collectively called the **stringent response**. Most remarkably, it down-regulates rRNA synthesis. This in turn will cause autogenous down-regulation of r-protein operon transcription. Another mechanism of posttranscriptional regulation is the signal-stimulated synthesis of antisense RNA. Concomitant **RNA duplex** formation will protect the double-stranded stretch of the mRNA from translation.

Chapter 13: Phage strategies

Answers to multiple-choice questions

1 (e); 2 (d); 3 (a); 4 (b); 5 (a); 6 (a); 7 (a); 8 (b); 9 (b); 10 (d).

Answers to concept questions

Question 1
Because assembly of phages is not controlled by enzymes, simultaneous expression of all phage genes would not allow timely synthesis of the progeny genetic element (double-stranded DNA) and the self assembly of the individual parts. In addition, it would obliterate the need for the lambda cI–Cro pair of competitive repressors; lysogeny would be impossible.

Question 2
The repressor, cI, of the first infecting lambda phage will bind immediately to the operator sites in the DNA of a second infecting phage genome thereby preventing the expression of the *cro* and *N* genes both of which are necessary for the lytic cycle. Hence, reinfection does not yield a virulent phenotype.

Answers to problems

Problem 1(a)
A mutation creating a protease-resistant λ cII protein will lead to lysogeny. Everything in λ phage infection goes normally until the N protein (expressed leftwards from P_L), antiterminates gene expression beyond the termination sites downstream of the *N* gene (on the left), leading to the expression of the cIII protein, and downstream of the *cro* gene (on the right of P_R), leading to the expression of the cII protein. Because the mutant λ cII protein is protease resistant, its action does not depend on the

status of the infected cell (healthy or sick) and the cII protein has very little to do to maintain the integrity of the λ cII protein. The resulting high concentration of λ cII-protein facilitates expression from the P_{RE} and P_I promoters leading to the synthesis of the cI (repressor) protein as well as the *cro* antisense RNA and integrase (for the incorporation of phage DNA), respectively. The repressor protein will occupy the O_R and O_L operators thereby autogenously facilitating its own synthesis (expression from P_{RM}) and preventing the synthesis of more N protein and Cro protein.

Problem 1(b)

A mutation of λ O_{R2} that prevents protein binding will result in lysis. A mutation in λ O_{R2} will prevent both the Cro protein and repressor from binding here. *cro* gene expression does not need an inducer, hence the Cro protein will be produced at high levels. The Cro protein will also bind to O_{R3} since its binding is not cooperatively. Because of this and the fact that repressor gene expression needs auto-induction by its own product, repressor is not being made. Cro will eventually shut off the expression of all early genes and induces intermediate and late gene expression through inducing expression from P_Q. Taken together, Cro wins out over the repressor.

Problem 1(c)

A mutation that inactivates the λ N gene results in inactivation and degradation. The N gene product is an antiterminator that is needed to allow gene expression beyond the N and *cro* genes. The result is the synthesis of a lot of faulty N protein and of Cro protein. There is no way that the phage can go into either the lysogeny or the lytic cycle and the phage DNA will, eventually, be degraded.

Problem 1(d)

A mutation in the gene encoding the λ cI protein results in quick lysis. If no functional repressor protein is produced—which means one that is unable to bind to the O_R and O_L operators, the result is that Cro wins out over repressor because there is none to compete with in the first place.

Problem 2

Lambda phage usually enters the lysogenic phase upon *E. coli* infection. Lysogeny is maintained autogenously by lambda repressor, cI, which must bind to the operator region. If a compound is a carcinogen it will most likely be a mutagen and, hence, may affect repressor expression (by causing a missense or nonsense mutation in the *cI* gene) or binding of repressor to the operator region (O_{R1} and O_{R2}) of the P_{RM} / P_R promoter region (by causing a mutation in operator sequence). By using the plaque assay, which is very cheap, appearance of plaques can be correlated with the mutagenic/carcinogenic action of the tested chemical.

232

Answer to summary worksheet

Viruses can cause lytic, persistent or latent infections of host cells that either destroy the invaded host immediately or they can place the cell's genetic and metabolic machinery under immediate (persistent) or future (latency) control of the virus for a longer period of time. Bacterial viruses or **bacteriophages** attack solely bacteria, hence they prey on single cells. Most phages can establish a latent infection in their hosts, called **lysogeny**, which allows propagation and survival of both bacterium and phage. The lysogenic nature of phage–bacterial interactions is host specific, which means that a given virus may be always lytic (virulent) in one bacterium while it may be lysogenic and capable of conversion to the lytic cycle (temperate) in another bacterial species. This requires that the viral genome is protected from destruction by host restriction systems during the latent phase and that it provides information allowing for differential regulation of viral replication and expression. This is needed independent of whether or not the lytic phase was preceded by lysogeny because phages are usually complex structures and assembly is solely controlled by the specificity and affinity of the viral parts for one another.

The best studied temperate phages are the so-called 'T-even' phages and lambda phage. Upon infection (injection of its double-stranded DNA), lambda phage DNA (λDNA) is circularized and **integrase** facilitates integration of λ**DNA** into the bacterial genome in a single reciprocal crossover event. Whether lambda remains lysogenic or goes into the lytic phase is ultimately determined by the concentration of two proteins, the **Cro protein** and the λ **repressor** (or **cI** protein), which compete for binding at the same transcriptional **operators** located between oppositely directed promoters, P_{RM} ('repressor maintenance'; leftward) and P_R ('rightward'). P_{RM} initiates transcription of the λ **repressor** gene, *cI*, whereas the rightward transcription initiated from P_R will yield the *cro* transcript. λ **repressor** binds cooperatively and with high affinity operator regions O_{R1} and O_{R2} which will allow positive autogenous regulation of *cI* gene expression from P_{RM}. In addition, if λ **repressor** can bind to the O_{R1} and O_{R2} regions, it will prevent binding of **Cro protein** to the O_{R3} and O_{R2} operator regions as well as the initiation of *cro* gene transcription by **DNA polymerase** from P_R (which overlaps with O_{R1}). Conversely, Cro protein binds with higher affinity for operator regions O_{R3} and O_{R2} and prevents the binding of λ **repressor** to the O_{R1} and O_{R2} operator regions that is required for expression from P_{RM} (which overlaps with O_{R3}). In essence, a high cI to Cro protein ratio will inhibit transcription from P_R and thus maintain **lysogeny** while the degradation of λ **repressor** will allow conversion into the **lytic cycle**. The integrity of λ **repressor** is maintained by functional cII and cIII proteins, whose synthesis is dependent on successful transcription from the 'rightward' and 'leftward' promoters, P_R and P_L.

233

The complex regulation of the lytic cycle is mostly accomplished by antitermination of transcription (read-through of DNA polymerase) which establishes two characteristic phases. The early phase accounts for expression of Cro protein and antiterminator protein, pN, which is needed for the 'delayed early' transcription of genes encoding factors essential for recombination and replication of λDNA as well as antiterminator protein, **pQ**. The latter will facilitate initiation and cause read-through of transcription from late phase promoter P_R. These late **mRNAs** will be then translated into head and tail parts of the virus.

Chapter 14: The replicon

Answers to multiple-choice questions

1 (c); 2 (a,c,d); 3 (b,d,e); 4 (a); 5 (c,d); 6 (b); 7 (c); 8 (a,b,d); 9 (a).

Answers to concept questions

Question 1

The Meselson–Stahl experiment in 1958 proved that replication is semiconservatively. Two hypotheses were, in fact, verified by this experiment: (i) replication requires separation of the two DNA strands (melting, denaturation) and (ii) the new DNA exists in two replica synthesized using the parental strands as templates. Experimentally, density gradient centrifugation was used which allowed the distinction of ^{13}C- and ^{15}N-labeled (heavy) DNA from ^{12}C- and ^{14}N-labeled (light) DNA. After the first generation time the DNA was of intermediate density. The importance of this discovery for our understanding of inheritance was that it provided a mechanism for faithful replication.

Question 2

Look at Figure 14.11 for D-loop replication. L-strand replication begins earlier and from an origin that is separate from that of H-strand replication.

Question 3

Linear replicons can be established at the end of linear chromosomes in the following ways:
(a) Short repetitive sequences at the ends of chromosomes allow unprecise replication accomplished by RNA polymerase priming ('telomerase').
(b) Terminal protein that is covalently linked to the template (5' end) strand provides a free nucleotide 3' end.

(c) Circularization of the double-stranded DNA and nicking generates a free 3'-OH end that can be extended, usually by rolling circle replication.

Question 4

Both are similar to the rolling circle replication in that dedicated proteins (TraY/I) nick the F plasmid at *oriT* thereby creating free 5' and 3' ends. Multiple proteins contribute to the unwinding, displacement and complete separation of the remaining F circle and the single strand, the resynthesis of the double strand in the donor, the protection and transport (conjugation) of the single strand to the recipient and the resynthesis of the duplex in the recipient. In Hfr conjugation, the bacterial chromosome is nicked at *oriT* and transconjugation of donor chromosomal DNA occurs as long as the conjugation pilus stays attached to the recipient. If arrested during pilus connection, the process could theoretically lead to the horizontal transfer of the entire chromosome.

Question 5

Multiple mechanisms contribute to the survival of plasmids in their host cells. The simplest version is multi-copy existence which ensures that both daughter cells contain at least one plasmid after binary fission is completed. Other mechanisms are killer–antidote systems, partition systems that connect plasmid replication with separation towards the two poles of a dividing cell and the location of stress tolerance and catabolic loci on plasmids. Plasmid incompatibility is achieved, for instance, by the 'competition' of different origins for initiation of replication and by countertranscript concentration (e.g., RNA I is antisense RNA to primer RNA).

Answers to problems

Problem 1

The single copy control is accomplished by 'measuring' the concentration of origins of replication per cell. Under optimal nutritional conditions, cell growth is fast, thus diluting the concentration of origin-sensing repressors and allowing continued initiation of replication. This enables the cell to initiate replication before the end of the preceding division cycle(s).

Problem 2

A reliable way to transfer a mutagenized gene between species is triparental mating where the recombinant host of the mutant gene construct is the donor, the wild-type to be mutagenized is the recipient and a third organism is the helper. The helper strain carries a plasmid which contains loci that allow selective initiation of the expression of the *tra* region in the recombinant donor plasmid. Both plasmids should have different origins of replication that are not recognized by the recipient's replication machinery. This system allows safe delivery of a suicide vector construct.

Answers to summary worksheet

The entire genome of a cell, **chromosomal** as well as **extrachromosomal** DNA, has to be replicated in order to maintain its hereditary information. In addition, the regulation of this process must ensure that the replication of the DNA occurs only **once** per cell cycle which is completed with the generation of **two** daughter cells. Surprisingly, this is accomplished by a very simple but effective mechanism called **semiconservative** replication. This means that there is no free or **template**-independent synthesis of DNA; new always pairs with old (as shown by the Meselson and Stahl experiment). In other words, genetic/hereditary information is either maintained (by **replication**), acquired (by **DNA uptake**) or created by alteration of existing information (**mutation** and **recombination**) but not created **per se**. Mechanistically, the process of replication is fairly complicated; you have to remember that the nucleic acid duplex is wound up into a **helix** which is organized at various further levels into **chromatin** structures. All these structures have to be undone in an orderly manner in order to 'melt' the two DNA strands and replicate, and the process must also allow the fast and faithful reorganization of the two replica duplex structures.

The structural unit of replication is the **replicon** which is defined by a site of replication initiation, the **origin**, and its termination. By recognizing the **cis**-acting nature of an origin, one can say that any DNA that is not separated from an origin by a termination site belongs to the same replicon as the origin. The replicon extends on the DNA molecule either unidirectionally or bidirectionally; most replicons are established simultaneously from overlapping origins in the duplex, hence they are **bidirectional**. In the case of prokaryotic circular DNA such as bacterial and plasmid DNAs, the most common mechanism of replication is the so-called **theta** replication in which the two replication forks move away and towards one another at the same time, hence the original and the replica do not go through **single-stranded** intermediates. Termination is accomplished by termination sites which are located downstream of the **fork meeting point**. Replication of circular DNA with a single-stranded intermediate is the **rolling circle** replication (example: circularized phage DNA). This requires a single-strand break at the initiation site and allows **amplification** of the original replicon into linear, **single-stranded**, single- or multi-copy replicas which need protection from degradation (by SSB). All eukaryotic, some prokaryotic (spirochetes) and some plasmid DNAs are linear. Because of this linear nature and because of the much larger size of the eukaryotic genome, replication of eukaryotic DNA involves many **replicons** of which only about 15% are active at any given time during **S**-phase. While this constitutes a regulatory problem in terms of ensuring that all replicons fire only once, the real challenge comes with the initiation and termination of individual replicons. Termination is usually achieved by the **merger** of bidirectional replicons, a process that is still not completely understood. The complicated mechanism of priming at the 3' ends of the linear DNAs is better understood. In phage DNA, polymerization is primed by a **terminal protein** (covalently linked to the template 5' end) which provides an extendable **cytidine** nucleotide (cytidylate) 3'-OH group. In eukaryotic chromosomes, this is accomplished by an RNA-dependent **DNA** polymerase (called **telomerase**) that generates **cDNA** from an RNA primer, thus it is a reverse transcriptase (see Chapter 26).

The presence of multiple replicons in eukaryotic linear DNA constitutes another difference between prokaryotic and eukaryotic cells in that the unit of replication does not coincide with the unit of segregation. Thus, the mechanisms reponsible for limiting the activity of a replicon in eukaryotic linear DNA to only once between **two cell cycles** must be different from those in prokaryotic circular DNA. This is addressed in detail in Chapters 26 and 15.

Chapter 15: DNA replication

Answers to multiple-choice questions

1 (b,c); 2 (a,b,d,e); 3 (a); 4 (a,c); 5 (a,c); 6 (a); 7 (c); 8 (a); 9 (b,d); 10 (b).

Answers to concept questions

Question 1

DnaA (which binds coordinatively at the 9 bp repeats, then melts the 13 bp repeats), DnaB (helicase) and DnaC interact with prokaryotic origins of replication before polymerase III is recruited. Since replication is semidiscontinuous, lagging strand replication requires multiple initiation of replication which is accomplished by the primosome (DnaG primase plus various protein factors dependent on the system).

Question 2

'Parental' DNA is usually methylated in a species-specific manner. After replication, the two template–replica duplexes are hemimethylated. Because hemimethylated DNA shows higher affinity for a membrane-associated receptor than for DnaA, hemimethylated DNA cannot be replicated. This prevents premature replication.

Question 3

The problem with the lagging strand is that the direction of replication (5' to 3') is opposite to the direction of fork movement. This requires semicontinuous replication in that it has to be primed multiple times which is accomplished by the primosome and the asymmetry of DNA polymerase III. The latter is asymmetrical and consists of subunits that allow for pulling the lagging strand in the direction of fork movement and simultaneous priming and primer extension (DNA synthesis) in the direction opposite of fork movement. Refer to Figures 15.15 and 15.16. Eukaryotic lagging strand synthesis is similarly accomplished.

Question 4

Because DNA polymerase has proofreading function, faulty bases will be recognized, excised by the exonuclease activity of polymerase and filled in by the correct bases. This recognition and repair occurs only as long as the DNA stays hemimethylated.

Question 5

Replication from the φX type origin requires the participation of additional proteins, the Pri proteins, which direct the primosome to the site of primer synthesis (at the primosome assembly site, *pas*). The primosome of the *oriC*-type origin consists only of DnaG primase.

Answers to problems

Problem 1

Experiments are outlined in Figures 15.25 and 15.26. Replication products from cells with and without a permeabilized nuclear membrane are analyzed for DNA density following the experiment of Meselson and Stahl. Permeabilization of the membrane simulates the nuclear breakdown that occurs during mitosis.

Problem 2

This situation requires that the initiation of a new round of replication occurs before the completion of the existing round, giving rise to multiple interconnected replicons. This will cut down the time required for binary fission to the time taken for cytokinesis. For a cell like *E. coli*, this time is approximately 20 min (the time required for replication is about 40 min) as compared with 60–70 min needed for binary fission under poor conditions. The main reason why the replication machinery involved in multiple genome copy replication does not collide is that the process is directed (5' to 3') and that *E. coli* has a circular genome. Directedness of DNA modification and polymerization is crucially important also for prevention of collision between the replication, transcription and translation machineries.

Answers to summary worksheet

While the replicon is the **structural** unit of replication and, hence, clearly defined, the replication process involves various enzyme complexes that form dynamically as they are needed. Four processes are to be considered: (i) initiation of replication at the **melting** by a proteinaceous initiation complex; (ii) repeated initiation of replication at the lagging strand by the **primosome**; (iii) elongation (unwinding or the parental strands and synthesis of the daughter strands) by the **replisome** and (iv) termination of the process at **termination sites** or by replicon **fusion**. The key step of initiation is the **melting** and **unwinding** of the DNA duplex at the origin which is accomplished by the DnaA and DnaB/DnaC proteins, respectively. While the **DnaA** and **DnaC** proteins are specific to the initiation process at the origin, the **DnaB** helicase is also part of the **primosome** and **replisome** complexes. In fact, it constitutes the minimum primosome because in *E. coli oriC*-type replicons DnaB is sufficient to interact with and locate the DnaG **primase** to the initiation sites of **Okasaki** fragment synthesis on the **lagging** strand. In other types of replicons such as phage φX, additional priming proteins (e.g., PriABC) assist DnaG **primase** with the recognition. **Primosome** assembly and disassembly also happens in linear eukaryotic replicons, hence it is a periodical process that generally occurs multiple times at the lagging strand near the replication fork in an active replicon.

Elongation of the primer, or daughter strand synthesis along the parental template is facilitated by DNA-dependent DNA polymerases and accessory enzymes. First, a reduction in DNA **supercoiling** is necessary which is accomplished by topoisomerases. Second, DnaB **helicase** (as a member of the primosome) unwinds the helix, melts the DNA in the fork and facilitates synthesis by interacting with **DNA polymerase**. While there is similarity between some subunits of eukaryotic and prokaryotic polymerases, the priming and organization of the elongation complex are significantly different in prokaryotes and eukaryotes. The DNA replicase in bacteria is an asymmetric dimer of **DNA polymerase III**, where the asymmetry stems from additional subunits that allow simultaneous synthesis of leading and lagging strands. This is remarkable because lagging strand synthesis occurs in a direction **opposite** to the movement of the fork. Prokaryotic **DNA polymerase I** exercises its **exonuclease** activity (RNA primer removal) and it extends the previous (free 3' end) to the new (5' end) Okazaki fragment before the phosphodiester bond is introduced by DNA **ligase**. In contrast, RNA primer removal and replacement by DNA synthesis are carried out by two different enzymes, the **exonuclease** MF1 and the DNA **replicase** that catalyzes nucleotide polymerization of the leading strand and the Okazaki fragments. All this contributes to the approximately ten times **higher** speed of prokaryotic compared with eukaryotic replication. Last, another important feature of replication is that the newly synthesized daughter strands initially lack any methylation, hence both duplexes are **hemimethylated**. Recognition of the DNA methylation state contributes to the regulation of various mechanisms that follow DNA synthesis such as the **repair** of unfaithful replication and mutations, reinitiation of the next replication cycle and segregation (in prokaryotes). In addition, it renders the DNA vulnerable to the cell's own **restriction** system.

Chapter 16: Restriction and repair

Answers to multiple-choice questions

1 (a,b,d); 2 (a); 3 (b); 4 (c); 5 (b); 6 (b,d); 7 (b); 8 (a).

Answers to concept questions

Question 1

Most endonucleases are 'restricted' in their action to specific sites (with a particular nucleotide sequence or methylation state). In addition, restriction endonucleases are designed to cleave foreign DNA such as infecting phage DNA. Phages are specific to their hosts, because in other organisms their DNA would be a sure target of the cell's restriction system. In that sense one can say that endonucleases restrict phages to their hosts.

Question 2

As already discussed in Chapter 15, replicated DNA is hemimethylated. This feature also applies to repair DNA synthesis (by polymerase I), hence as long as methylation does not occur (according to the species-specific pattern), it is possible to distinguish between the parental and the replica strand. In that sense, the methylation state of DNA is crucial to the maintenance of species genotypes.

Question 3

Refer to Figure 16.13 which schematically shows that UvrA localizes the damage and facilitates binding of UvrB. UvrB recruits UvrC which is an endonuclease capable of nicking the DNA upstream and downstream of the site of injury. Unwinding by UvrD allows removal of the damaged DNA segment and filling-in of complementary DNA by polymerase I.

Question 4

Again, the *mut* system of DNA mismatch repair recognizes the methylation state of the DNA. In addition, there are dedicated enzyme systems that replace mismatch base pairs preferentially (e.g., it is more likely that G•T will be changed to G•C than to A•T).

Question 5

The RecA protein recognizes damage to DNA and triggers the autocatalytic cleavage of the LexA protein. The inactivation of LexA removes the repressor of operons that contain genes encoding crucial repair and defense proteins such as *recA*, *uvrAB* and *umuC*. Because *lexA* expression is regulated autogenously, the SOS response simultaneously produces the necessary repair functions and the product that will end the response and cause the return to normal: repression of these operons by the LexA protein.

Answers to problems

Problem 1

The phenomenon of multiple copy identity has been observed but the mechanisms are not known in much detail. In order to maintain multiple, nearly identical copies, the additional copies must some how be compared with a 'master' copy which may be close to the ancestral sequence. This comparison between copies could be facilitated by recombinational strand exchange such as that which occurs during recombinational repair of heteroduplex DNA or gene conversion. In addition, strongly enforced mismatch repair would reduce wobbling and drift.

Problem 2

Once a lysogenic bacteriophage has incorporated into the host genome, it constitutes a prophage. Prophage DNA differs from the host DNA in its methylation state as long as it has not been replicated with the host's genome. Once replicated, prophage DNA is subject to host-specific DNA modification and hence is no longer distinguishable. This and the fact that phages can incorporate into host DNA at both specific (*att*) and nonspecific sites makes the utilization of repair systems for this task unlikely. With some luck it might be possible to recognize prophage incorporation with Southern RFLP analysis. Potentially the quickest experiment would be based upon the fact that prophages go into the lytic cycle upon exposure of the lysogens (bacteria with prophages) to environmental stress signals such as UV light. This would allow the design of a 'plaque induction assay'.

Answers to summary worksheet

Because they are haploid, bacteria are under tremendous stress to defend the integrity of their DNA against contamination by **foreign** DNA as well as against loss of crucial hereditary information caused by faulty replication or mutations. To this end, bacteria contain specific **endonucleases** that recognize characteristic sequences as well as the state of methylation to distinguish between **self** and **foreign** DNA. These restriction systems have to recognize foreign DNAs before they are replicated because host **methylases** will remethylate any DNA in a manner specific to the cell that presents the appropriate recognition sequences.

The methylation state of the DNA after replication also provides crucial information for systems involved in the repair of injured DNA. The **hemimethylation** allows the distinction between the **parental** and the **daughter** strands and, hence, correction of a mismatch or injury by using the parental strand as the **template**. Excision repair, which can be very short, short or long, involves the activity of the **UvrABC** protein complex (which has recognition and endonuclease activity), **DNA polymerase I** (which has exonuclease activity and synthesis of DNA replacement) and **DNA ligase** (which seals nicks). Many mutator loci have been identified whose products confer the ability to carry out **mismatch** repair. These *mut* gene products are able to specifically replace individual mismatched nucleotides such as T from G•T and C•T mismatches.

Repair after replication and remethylation may involve using the healthy daughter duplex for repair of the injured or faulty daughter duplex. In addition to endo- and exonuclease activities, this repair involves the **RecA** protein which facilitates single strand invasion and strand displacement in recombination. For this reason, retrieval systems are also known as **recombinational** repair systems. If the RecA protein is activated for repair, it will trigger the induction of expression of genes that are usually repressed by the **LexA** protein. Many of these products have repair functions and the response that can occur as fast as minutes after the DNA injury is known as the **SOS** response.

Chapter 17: Recombination

Answers to multiple-choice questions

1 (b,e); 2 (a,b,d,e); 3 (b); 4 (b); 5 (c,d); 6 (a); 7 (b,c,d); 8 (b,e).

Answers to concept questions

Question 1

If homologous DNA strands are nicked (the strands belong to two different duplexes!), broken strand exchange between the duplexes and branchpoint migration produce a region of heteroduplex DNAs (Figure 17.2). To resolve the cointegrate, a second nick has to occur which can be in the same or the complementary DNA strands. Nicks in the same strand lead to parental (patch) recombinants while nicks in the complementary strand create (splice) recombinants made up of heteroduplex DNA (Figure 17.4).

Question 2

While there is a chance that DNA and thus information will be lost as a consequence of DNA double breaks, loss can be prevented in the presence of a homologous duplex. Creation of a free single-stranded 3' end at the DNA injury as a first step is followed by single strand invasion and strand displacement in the healthy duplex. The displaced strand is then used as template for repair synthesis of the other strand of the injured DNA duplex (Figure 17.5). This creates two oppositely migrating branchpoints that constitute a double crossover (= a double Holliday junction). It is crucial that the branchpoints move into healthy DNA duplex regions where they can be resolved by nicks.

Question 3

Branchpoint migration during single-strand assimilation is facilitated by a hexamer of the RecA protein. Hydrolysis of ATP is required for dissociation of the RecA complex from DNA. RecA can also facilitate branchpoint movement in duplex recombination if it involves a significantly extended single-strand 3' end as found in double-strand recombinational repair. Brachpoint migration in cointegrated DNA duplexes requires the action of the RuvAB protein complex. The complex recognizes crossover points and binds and disconnects from the DNA with hydrolysis of ATP thereby moving the branchpoint into the duplex DNA which extends the heteroduplex region.

Question 4

Gene conversion is the process of changing the DNA sequence of one strand in a mismatched or heteroduplex to complementarity to the other strand. In that sense it is a result of unreciprocal transfer of information/loci but not necessarily crossing over. Also single-strand assimilation can lead to segments of noncomplementary heteroduplex DNA. As a result of gene conversion, the segregation of loci differs from the expected parental segregation ratio (Fig. 17.13).

Question 5

Only gyrase, a topoisomerase Type 2 (breaking and religating both strands), is capable of introducing negative supercoils. It hydrolyzes ATP to gain the energy necessary to catalyze the reaction from a state of lower free energy into a state of higher free energy. The linking number changes by two (Fig. 17.17).

Answers to problems

Problem 1

A and A' contain a recognition site for *Eco*RI which is abolished in a and a' by base substitution:

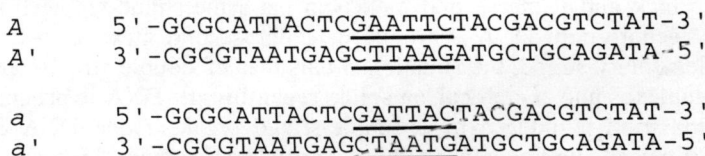

```
A     5'-GCGCATTACTCGAATTCTACGACGTCTAT-3'
A'    3'-CGCGTAATGAGCTTAAGATGCTGCAGATA-5'


a     5'-GCGCATTACTCGATTACTACGACGTCTAT-3'
a'    3'-CGCGTAATGAGCTAATGATGCTGCAGATA-5'
```

If the locus of interest is near the crossover point, crossing over could appear as a nonreciprocal event. Double-strand break repair and meiotic recombination generate two crossover points with two heteroduplex regions in between. Which of the alleles (A and A' or a and a') is maintained in the progeny depends on both branchpoint movement and to which of the parental strands the sequence adjustment occurs (mismatch repair). Therefore, if the *Eco*RI site is within the locus of interest, a 2:2 ratio would indicate a regular segregation; however, a 3:1 or 1:3 ratio means that one of the site mutations has been repaired to the original sequence or vice versa. Check haploid yeast cells using RFLP.

Problem 2

The isomeric Holliday structure (in which the lengths of the protruding 'arms' are equal) is the result of homologous recombination. Nicks in the same strands where the first nick occurred lead to patch recombinants (recombinant patches within parental duplexes). Nicks in the other strands lead to splice recombinants (all four strands cut at crossover; DNA duplexes from different parental double strands are connected via heteroduplex DNA). Consult Figures 17.3 and 17.4.

Answers to summary worksheet

While **mutations** allow small changes in the genome that may or may not manifest themselves in the phenotype, **recombination** generates significant changes and can be viewed as the motor of evolution. Recombination involves the exchange of large segments of DNA within or between **chromosomes**, thus it occurs usually between two DNA duplexes. However, DNA duplexes can also assimilate single-stranded DNA which is mediated by the **RecA** protein. The reciprocal exchange of **homologous** (sharing common ancestor) highly similar DNA sequences is called homologous recombination and this kind of recombination plays an important role in recombinational repair. Naturally, eukaryotic chromosomes form synaptonemal complexes during **meiosis**, a stage at which the **reciprocal** crossing-over between the duplexes occurs. This becomes visible in form of the **chiasmata**, places at which the duplexes are held together during chromosome separation.

Recombination can be initiated by double-strand breaks (both strands in one duplex) or breaks in homologous strands of two duplexes. The latter leads to **single-strand** exchange in that the free **3' ends** invade the other duplex to pair with the unbroken strands. This creates a crossover point or a 'Holliday junction'. The Holliday junction is resolved by a second nick and if the second nick is in the same strand as the first nick, **patch recombinant** DNA is produced. If the second nick is made in the complementary strand, the situation equals that of double-strand breaks in both duplexes and reciprocal or **splice recombinant** DNA is produced. If recombination is induced by a double-strand break in one DNA duplex, free 3'-OH ends will be generated by **exonuclease** activity. In bacteria, this may be facilitated by the RecBCD complex. The following single-strand invasion of the healthy duplex displaces one strand comparable to mitochondrial D-loop replication. The displaced strand will serve as **template** for synthesis of the other strand in the injured duplex. At this point a double crossover is established and the branchpoint moves toward healthy DNA. Branchpoint movement is powered and carried out by the **RuvAB** protein complex.

The incorporation of phage DNA into the host bacterial genome also employs reciprocal recombination as described for meiotic crossing-over or double-strand break-induced recombinational repair. However, phage DNA incorporation is an example of **site-directed** recombination and it occurs rarely at sites in the bacterial genome different from phage DNA attachment sites (*att* sites). In addition, this process usually does not involve any DNA synthesis and the reciprocal strand exchange is not facilitated by the RecA protein. Instead, phage and host factors assemble to form the so-called '**intasome**.'

If the crossing-over between alleles is nonreciprocal or the single strand assimilation is of nonhomologous nature, the segregation of traits will not follow Mendelian rules. This effect is known as **gene conversion**.

Crucial to all processes of DNA modification is the ability of the cell to regulate the tension produced by the helix structure and the degree of coiling of the DNA duplex. This topological state of the DNA is described by the **linking number** and a change in the linking number requires the temporary breakage of at least one strand in the duplex. The enzymes able to break and religate the DNA in order to relax or increase **supercoiling** are called **topoisomerase** and they are classified by their ability to break and religate one or both strands (Topo I and Topo II, respectively).

Chapter 18: Transposons

Answers to multiple-choice questions

1 (b); 2 (b,d); 3 (a,b,c); 4 (a,c,d); 5 (b,c,e); 6 (b); 7 (a,b,d); 8 (a); 9 (b); 10 (a,c,d); 11 (a,b,c,d); 12 (b,c,d)

Answers to concept questions

Question 1

Transposons can cause deletion, duplication, or insertion of host sequences when they move. In addition, transposons can be acted upon by host recombination systems to cause genome rearrangement.

Question 2

The target site is duplicated because a staggered cut is made prior to transposon insertion and filled in after transposon insertion.

Question 3

A composite transposon has an IS element on each end.

Question 4

First, a staggered nick is made at the target site and the transposon is cut out. Next, the transposon is joined to the target site. Finally, the single-stranded regions on both sides of the insertion site are filled in.

Question 5

Recombination between direct repeats deletes DNA sequences between the repeats. Recombination between inverted repeats inverts the DNA between the repeats.

Question 6

A cointegrant is formed during replicative transposition. It has two copies of the transposon in the same orientation and separated by the original replicons.

Question 7

As soon as the transposase has been translated, it could bind tightly to DNA and thus be prevented from diffusing to other elements elsewhere in the genome. Another hypothesis proposes that free transposase has a short half life, but the protein is stabilized when it binds DNA. Because the unbound form is unstable, it would not diffuse to bind to other sites.

Question 8

Hybrid dysgenesis is a set of chromosome abnormalities that occur in the offspring when certain strains of *Drosophila melanogaster* are crossed. These abnormalities result from the activation of P element transposons that are present in the chromosomes of one parent but not the other. The P elements become active if the mother does not have P elements and therefore produces eggs that lack a P element-encoded repressor of transposition.

Answers to problems

Problem 1

TnpR has two functions: resolvase and transcriptional repression. When the repressor is mutated, expression of *tnpA* (transposase) is higher, leading to more frequent transposition. How could you prove that TnpR acts as a transcriptional repressor (hint: try using reporter genes)? What mutation would render a promoter immune to a repressor?

Problem 2

Which of these mutants are capable of transposition? Recall that expression of too much transposase is deleterious to the cell.

Answers to summary worksheet

Transposable elements are DNA sequences that can **move** to new locations in the genome. This can affect the genome in a variety of ways: they can cause gene **rearrangement**, they can **inactivate** genes by inserting into them, and their promoters can affect the expression of neighboring genes.

The simplest transposable elements are called **IS** elements. These contain short **inverted** terminal repeats surrounding a gene encoding **transposase**, the enzyme responsible for transposition. Upon integration into a new site, transposable elements always create a **direct** repeat of the target site.

Composite transposable elements have two **IS** elements flanking a gene encoding **antibiotic** resistance. Some elements move by **replicative** transposition, in which a copy of the element is left behind at the original site. This mechanism produces a **cointegrant** intermediate that contains two copies of the element. The enzyme **resolvase** then acts to cause homologous recombination between the two copies of the element. Other elements use **nonreplicative** transposition, in which the element moves to a new site but is not copied. This type of transposition requires **breakage** and **reunion** of the **target** site (the **donor** site remains broken).

Transposition of Tn10 **is** tightly controlled by keeping the level of transposase very **low**. Transposase is synthesized from the *IN* RNA made by the P_{IN} promoter of **IS10R**. The P_{OUT} promoter makes the *OUT* RNA, which is **complementary** to part of the *IN* RNA and which is **more** abundant than the *IN* RNA. *OUT* RNA controls the expression of transposase by hybridizing to the *IN* RNA and preventing its **translation**. Tn10 transposition is also regulated by DNA **methylation**. First, **methylation** of a site in P_{IN} prevents **transcription**. Also, **methylation** of a site in the right-hand inverted repeat prevents **transposase** binding.

Transposable elements were first identified by the pioneering work of Barbara McClintock, who studied variegation of maize (corn) kernel phenotypes. She defined maize transposons called **controlling** elements. When these elements move they can cause chromosome **breakage**, resulting in acentric fragments. These fragments are lost the next time the cell goes through mitosis, leading to a change in **phenotype**. **Autonomous** elements such as *Ac* are competent to transpose while **nonautonomous** elements like *Ds* are not. These elements have suffered internal **deletions**, and can only transpose when an **autonomous** element is present in the genome. Transposable elements have also been identified in *Drosophila melanogaster* on the basis of **hybrid dysgenesis**. This phenomenon involves a variety of chromosome **aberrations** and is observed among the offspring when certain strains of flies are crossed. The aberrations are due to transposable elements known as **P elements** which are activated if the egg does not contain a protein that **represses** transposition.

Chapter 19: Retroviruses and retroposons

Answers to multiple-choice questions

1 (a,b,c,d); 2 (b,c); 3 (c); 4 (a,c,d); 5 (b,c); 6 (a,b); 7 (c); 8 (a,b,d,e,f); 9 (c,e); 10 (b,e)

Answers to concept questions

Question 1

Retrotransposons are mobile genetic elements that pass through an RNA intermediate when they transpose. Retroviruses are a subset of retrotransposons that form infectious particles containing an RNA genome.

Question 2

All structural proteins of a retrovirus are generated from a single primary transcript. The first protein made is Gag. To synthesize the Pol protein, the stop codon at the end of the Gag coding region must be suppressed or the ribosome must shift reading frames. Thus, Pol is made only as a Gag–Pol fusion protein, which is subsequently cleaved by a viral protease. The Env protein is made from an alternatively spliced mRNA (alternative splicing is discussed in Chapter 30).

Question 3

After Gag, Pol, and Env have been translated, the viral protease cleaves them each into two or three fragments, which are the active forms of the viral proteins.

Question 4

Refer to Figures 19.6 and 19.7.

(i)   The plus-strand RNA genome is brought into a host cell by a virus particle.

(ii) A host tRNA anneals to the viral RNA downstream of the U5 region.

(iii) Reverse transcriptase uses the tRNA as a primer to begin synthesis of a minus-strand DNA.

(iv) Reverse transcriptase stops when it reaches the end of the template, making the strong stop minus DNA.

(v)   The 5' end of the template RNA, corresponding to the R region, is digested by the RNAase H activity of reverse transcriptase (RNAase H degrades the RNA strand of a DNA–RNA helix).

(vi) The strong stop minus DNA makes the first jump to anneal to the R region at the 3' end of the viral RNA.

(vii) Reverse transcriptase extends the strong stop minus DNA to make a long minus-strand DNA that still contains the tRNA primer on its 5' end.

(viii) The tRNA primer is removed.

(ix) Most of the viral RNA is degraded.

(x) Plus-strand DNA synthesis is primed by remaining fragments of viral RNA to make the strong stop plus DNA. Note that this reaction is also catalyzed by reverse transcriptase, which is now using DNA as a template.

(xi) The strong stop plus DNA makes the second jump to anneal to the U5 region on the 3' end of the long minus-strand DNA.

(xii) The plus strand is extended until it is complete.

(xiii) The synthesis of the minus strand is completed by finishing the 3' end of the 5' U3 element.

Question 5

The target site is duplicated because retroviral integrase generates a staggered nick at the integration site. After insertion, the nicked sequences are filled in to create repeats. Transposase generates target site direct repeats in much the same way. The deletion of two nucleotides from each side of the viral genome does not cause loss of this information as there is another copy of each sequence at the opposite end of the genome.

Question 6

Retroviruses acquire a cellular gene when a deletion occurs between an integrated provirus and a neighboring cellular gene. Transcription from the proviral promoter generates a fusion mRNA that, after being spliced, could be packaged into a virus particle. Retroviral genes such as *pol* or *env* can be lost when the host DNA is acquired. Such a virus would be unable to replicate on its own but could replicate if a wild-type helper virus is present.

Question 7

Each end of the Ty element is a direct repeat called a delta or δ element. The Ty element has two open reading frames that have homology to retroviral *gag* and *pol* genes. Translation of the second reading frame requires a ribosomal frameshift, just as retroviral *pol* does. Ty elements lack an *env* gene, preventing the production of infectious particles.

Question 8

Engineer an artificial Ty element containing an intron and a marked δ element. Use the inducible *GAL* promoter to generate lots of mRNA from this Ty element, which leads to frequent transposition of this element into new sites in the genome. Examine the newly-inserted Ty elements. If they have the marked δ but have lost the intron, they have gone through an RNA intermediate.

Question 9

FB is flanked by inverted repeats while *copia* is surrounded by direct repeats.

Question 10

Members of the viral superfamily have an LTR, open reading frames encoding reverse transcriptase or integrase, and introns. Nonviral retrotransposons have none of these. Also, integration of viral retrotransposons generates a small (4–6 nucleotide) duplication at the target site while the nonviral elements generate a larger (7–21 nucleotide) duplication upon integration.

Question 11

Alu elements contain 14 nucleotides that are nearly identical to the origins of DNA replication of some viruses, suggesting that Alu elements might serve as replication origins. However, the number of Alu elements in the genome exceeds the estimated number of replication origins by tenfold.

Answers to problems

Problem 1

What enzymes are essential for viral replication? Are these enzymes needed by the host? What kind of molecules could you use to inhibit enzymes? (Hint: all enzymes need to bind their substrates in order to work. Substrate analogs, which differ slightly from the true substrates, might still be able to bind the enzyme.)

Problem 2

The frameshift occurs by a –1 slippage of the ribosome. It occurs when the ribosome contains the Asn tRNA (anticodon 5'-AUU-3') in the P-site and the Leu tRNA (anticodon 5'-UAA-3') in the A-site (refer to Fig. 8.6 for a review of ribosome translocation).

```
              Asn      Leu
               |        |
             U-U-A    A-A-U
    5'- A C A A A U  U U A U A G G G A G G -3'
```

Both tRNAs then slip backwards one nucleotide. Although their anticodons are no longer perfectly paired with the mRNA, each tRNA maintains two base pairs with the mRNA.

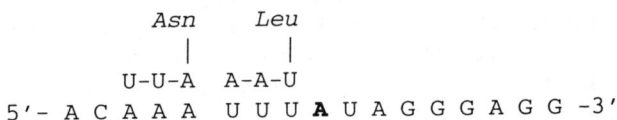

```
              Asn      Leu
               |        |
             U-U-A    A-A-U
    5'- A C A A A  U U U A U A G G G A G G -3'
```

This –1 slippage makes the next codon AUA instead of the UAG of the original reading frame. Thus, one nucleotide is decoded twice: UAA as a leucine codon in the original frame and AUA as an isoleucine codon in the shifted frame.

```
5'- A C A A A U U U  A U A  G G G A G G -3'
          A-A-U      U-A-U  C-C-C
            |          |      |
           Leu        Ile    Gly
```

The UUU sequence is critical for the –1 frameshift as any changes in it would destabilize the Leu tRNA, thereby terminating translation. The stem–loop structure immediately 3' of the frameshift region may be important for physically stopping the ribosome to allow it to slip into the –1 frame.

Answers to summary worksheet

The retroviral genome is made of **RNA**. When a retrovirus infects a cell, the genome is first copied into **DNA** by the enzyme **reverse transcriptase**. Like all DNA polymerases, this enzyme needs a **primer**. To begin this process, a host **tRNA** is used for this role. The **DNA** copy then **integrates** into the host genome to form a **provirus**.

Retroviruses contain three genes: *gag*, *pol* and *env*. Viral structural proteins are encoded by the *gag* and *env* genes while the *pol* gene encodes enzymes needed for viral replication including **reverse transcriptase**, **integrase** and **protease**. The **Gag** and **Pol** proteins are **translated** from the same mRNA. In some cases (where the two genes are in the same reading frame), translation of *pol* requires **readthrough** of the *gag* stop codon. If the two genes are in different reading frames, a ribosome **frameshift** is required to synthesize the *pol* protein. Both these processes are **inefficient**. Thus, while *gag* is made by the normal translation mechanism, *pol* can only be synthesized as a fusion protein. All viral proteins are **cleaved** by a viral **protease** to generate the mature viral proteins.

Retroviruses can acquire host genes such as *onc* genes. In the process, these **transducing** viruses lose essential viral genes and become replication-**defective**. A **helper** virus can provide normal viral functions, allowing these viruses to replicate.

Yeasts contain a transposable element called the **Ty** element, which is related to retroviruses. These elements move to new sites in the genome by passing through an **RNA** intermediate. The element contains open reading frames homologous to the retroviral *gag* and *pol* genes, although no *env* homolog is found. The *copia* element of *Drosophila* is quite similar.

Retrotransposons of the **viral** superfamily are similar to retroviruses in several ways. They are flanked by **long terminal repeats** and encode proteins with homology to **reverse transcriptase** or **integrase**. However, they do not form **infectious** particles. In contrast, members of the **nonviral** superfamily of retrotransposons **do not** encode any proteins and **are not** flanked by repeats. Both families are actively transcribed, but the viral elements are transcribed by RNA polymerase II and the nonviral elements by RNA polymerase III. Both families of elements transpose via **RNA**, and both are very **abundant** in mammalian genomes.

Chapter 20: DNA biotechnology

Answers to multiple-choice questions

1 (b,c,d,e); 2 (c); 3 (b); 4 (c); 5 (a,c,d).

Answers to concept questions

Question 1
(i)   reverse transcriptase to copy mRNA into single-stranded cDNA;
(ii) DNA polymerase to make single-stranded cDNA into double-stranded DNA;
(iii) S1 nuclease to create blunt ends on the double-stranded DNA;
(iv) restriction enzyme to cleave vector;
(v)   DNA ligase to insert cDNA into vector.

Question 2
A clone is any DNA fragment that has been inserted into a vector. Cloning allows production of large amounts of a DNA fragment to facilitate its analysis. A probe is a DNA fragment that is labeled and used in a hybridization experiment such as a Southern blot.

Question 3
A genomic clone contains all the intervening sequences present in a gene; these are not present in a cDNA clone.

Question 4
Both techniques are used to assay a mixed DNA sample for the presence of sequences that can hybridize to a particular probe. The dot blot is faster than a Southern blot (why?) and can analyze more samples at the same time. The advantage of a Southern blot is that this technique also reveals the size of the DNA fragments that hybridize to the probe. For example, Southern blots could detect gene rearrangements that dot blots would not.

Question 5
This is a trick question; PCR reactions commonly use incubations at three different temperatures. The first step is denaturation of the template DNA by heating to 95°C for one minute. In the second step the reaction is cooled to allow the primers to anneal. The precise temperature for this step will be determined by the 'melting temperature' of the primers. This is the highest temperature at which the primer can stably anneal to the template, and depends on the length and sequence of the primers (melting temperature increases with primer length and G+C content). Generally,

high melting temperatures are desirable as they help ensure that the primers anneal only to the correct target sequences in the template. The third step is carried out at the optimum temperature for the DNA polymerase. DNA polymerase isolated from thermophilic bacteria is often used in PCR reactions, allowing incubation at 74°C. A further advantage of thermostable DNA polymerase is that it retains its activity after short incubations at 95°C. Thus, multiple cycles can be performed without adding fresh DNA polymerase for each round. What would happen if the DNA polymerase were irreversibly denatured by the 95°C step?

Question 6

YACs require all the functional elements of natural chromosomes including a centromere, an origin of DNA replication and two telomeres. YACs can accommodate inserts of foreign DNA as large as 300 kb, which is much more DNA than plasmids or cosmids can hold. Large inserts are advantageous because they are more likely to contain intact genes, allow larger steps during chromosomal walks, and reduce the number of clones needed in a genomic clone library.

Answers to problems

Problem 1

Northwestern blots detect RNA binding proteins by incubating proteins immobilized on a membrane with a labeled RNA probe. What do you think a Southwestern blot detects?

Problem 2

If your hybridization probe is double stranded, you can also get a blank autoradiogram by forgetting to denature the probe before adding it to the hybridization mixture.

Problem 3

The complete digest will have four fragments; a partial digest will include these and six additional fragments (including uncut starting fragment). How many of these will contain an intact *YFG1* gene?

Answers to summary worksheet

Cloning relies on the use of many different DNA metabolism **enzymes** that have been isolated (predominantly) from prokaryotic cells. **Restriction** enzymes are required to cut DNA into manageable fragments. A variety of DNA **polymerases** are used including **reverse transcriptase**, which copies RNA into single-stranded DNA. DNA **ligase** is used to join restriction fragments together.

Cloning also requires DNA sequences from microbial organisms. These are called cloning **vectors**, and are used to propagate the cloned DNA. Plasmid vectors are derived from native bacterial plasmids and commonly carry genes conferring **antibiotic** resistance. Other vectors are made from

bacteriophages such as λ, which have been modified to carry foreign DNA. Very large fragments of foreign DNA can be cloned in vectors called **yeast artificial chromosomes**, which contain all the DNA sequences required for chromosome replication and segregation.

To clone a eukaryotic gene of interest, a **probe** must be made that will hybridize to the gene (this usually requires some knowledge of the DNA sequence of the gene). Also needed is a clone **library** that contains many thousands of plasmids (or phages or YACs), each of which carries a different fragment of eukaryotic DNA. When the probe is incubated with the clone library, it will **hybridize** only to those rare plasmids that contain a DNA fragment complementary to the probe.

Many powerful and sophisticated techniques have been developed to analyze cloned DNA. For example, DNA fragments present in very small amounts can be amplified using the **polymerase chain reaction**. Gene activity can be quantified by attaching a **reporter** gene such as **CAT** or *lacZ* to the promoter of the gene of interest.

Chapter 21: Genomes

Answers to multiple-choice questions

1 (c); 2 (a); 3 (a); 4 (b); 5 (c); 6 (b,c,d); 7 (d); 8 (d).

Answers to concept questions

Question 1

For higher eukaryotes, genome size does not correlate with the complexity of the organism. Closely related organisms can differ widely in DNA content. For example, some amphibians have one hundred times more DNA than other amphibians, yet are not substantially more complex creatures.

Question 2

$C_0t_{1/2}$ values are affected by genome size and by the types and amounts of repeated DNA present in the genome.

Question 3

When a poly(A)–poly(U) duplex is denatured and reannealed, any poly(A) strand can reanneal with any poly(U) strand, allowing very rapid reassociation. When MS2 DNA is denatured and reannealed, each fragment can only reanneal with its original opposite strand. Therefore, the reannealing reaction takes more time.

Question 4

Although both genomes are composed of unique sequence DNA, the *E. coli* genome is approximately 25 times larger. Therefore, each denatured strand of *E. coli* DNA takes longer to find its correct opposite strand for reannealing.

Question 5

The stringency of a hybridization reaction is determined by two factors: the temperature and the salt concentration. Higher temperatures require stronger base pairing, while higher salt concentrations stabilize weak base pairing. Thus, a high stringency hybridization is performed at high temperature and low salt concentration, while a low stringency reaction uses low temperature and high salt concentration.

Question 6

In a DNA-driven experiment, a small amount of radioactively labeled RNA is included during the reassociation of genomic DNA. Since the DNA is present in vast excess over the RNA, it 'drives' the hybridization of the RNA to completion. Similarly, an RNA-driven reaction is performed with excess RNA and limiting DNA, so that all DNA strands that can hybridize to an RNA will do so.

Question 7

$R_ot_{1/2}$ measures the reassociation of RNA.

Question 8

Begin by isolating nonrepetitive DNA. Denature and reassociate this DNA in the presence of excess mRNA, measuring how much of the DNA reanneals. Generally, only a small percentage of the DNA is found to anneal to RNA. Note that the maximum amount of DNA that can anneal to RNA is 50%, because only one strand of any DNA fragment will be complementary to an RNA. The other strand is identical to the RNA and cannot hybridize to it.

Question 9

From the $R_ot_{1/2}$ experiment in Figure 21.9, we see that 2.7% of sea urchin unique sequence DNA is transcribed into mRNA at the gastrula stage. From $C_ot_{1/2}$ analysis, we know that unique sequence DNA comprises 75% of the sea urchin genome, which contains 8.1×10^8 nucleotides. Thus the amount of unique sequence DNA that is transcribed into mRNA is $(0.027)(0.75)(8.1 \times 10^8) = 1.6 \times 10^7$ nucleotides. If the average gene is approximately 2000 nucleotides long, the transcribed unique sequence DNA represents about 8000 genes.

Answers to problems

Problem 1(b)

To perform $C_ot_{1/2}$ analysis: isolate genomic DNA, randomly shear this DNA into fragments, denature the DNA by heating, allow the DNA to reassociate, and follow reassociation by hyperchromicity (the change in UV light adsorption when single-stranded nucleic acids anneal to become double-stranded). To analyze the reassociation kinetics, plot the fraction of DNA reassociated versus $C_ot_{1/2}$; this analysis will detect repeated sequences (low $C_ot_{1/2}$) and unique sequences (high $C_ot_{1/2}$)

Problem 1(c)

What component of the eukaryotic genome reassociates like the prokaryotic genome?

Problem 1(d)

Isolate unique sequence fraction from the $C_ot_{1/2}$ experiment above. Isolate total cellular RNA. Hybridize the RNA to the unique sequence DNA under RNA excess conditions and follow reassociation kinetics as above.

Problem 2

Refer to Figure 21.4.

Problem 3

Are all the DNA duplexes formed in the fast-reannealing component perfect duplexes? Are they all the same? Are imperfect duplexes as stable as perfect duplexes? Are the duplexes formed in the slowly-reannealing component perfect?

Problem 4

Plan 1. Perform a reassociation experiment, removing aliquots after various times of incubation. When all the repeated DNA has reannealed, remove the double-stranded DNA from the reaction (you could use a column or filter that binds only double-stranded DNA). Now continue the reassociation reaction until all the unique DNA has reassociated. Make a clone library from this unique sequence DNA and use it for the sequencing project.

Plan 2. To sequence only expressed unique genes, hybridize the unique sequence DNA purified in Plan 1 to total cellular RNA. Let the reassociation go to completion. Now you remove the single-stranded DNA from the reaction. Only expressed, unique DNA is left in the form of DNA–RNA hybrids. Clone and sequence this DNA. (An alternative approach currently being used is to make cDNA from cellular RNA and sequence these cDNAs, which are called ESTs, for expressed sequence tags.)

Answers to summary worksheet

Genome size

The amount of DNA in the genome of an organism is called its **C value**. While lower eukaryotes generally have lower **C values** than higher eukaryotes, related organisms can differ widely in DNA content. Thus, DNA content does not always correlate with the **complexity** of an organism. An extreme example is found by comparing the genome sizes of flowering plants, where the largest genome contains **1000** times more DNA than the smallest. It is difficult to imagine that these plants differ so extensively in morphological complexity. In general, higher eukaryotes have much **more** DNA than expected.

Components of eukaryotic genomes

Reassociation kinetics is used to study the types of DNA found in a eukaryotic genome. The results are generally expressed in a $C_ot_{1/2}$ curve. In these studies, the genomic DNA is sheared randomly into fragments. The DNA is then heated to **dissociate** it into single strands. The reaction is then cooled to allow the single strands to **reassociate** and the amount of DNA reassociated at various times is measured. Reassociation depends on two factors: C_o, the initial DNA **concentration** and $t_{1/2}$, the **length of time** of the reactions. $C_ot_{1/2}$ describes the point in the reaction at which **50%** of the DNA has reassociated.

For a simple duplex of homopolymers, such as poly(A)–poly(U), $C_ot_{1/2}$ is **low** because any poly(A) strand can pair with any poly(U) strand. For DNA samples with more complex sequences, each strand must find its correct opposite strand, which increases the **time** required for reassociation. Therefore, $C_ot_{1/2}$ **increases** with genome size.

$C_ot_{1/2}$ analysis can also be used to determine genome **complexity**. The *E. coli* genome has **one** reassociation component which consists of **unique** sequence DNA. In striking contrast, eukaryotic genomes generally contain **three** components which reassociate independently, demonstrating that the organization of the eukaryotic genome is fundamentally different from that of the prokaryotic genome. The **fast** component comprises roughly **25%** of the genome. This component has a **low** $C_ot_{1/2}$ value, reassociates **more** quickly than the *E. coli* genome, and is composed of highly repeated sequences. The next component to reassociate is called the **intermediate** component, representing approximately **30%** of the genome. This component has a **low** $C_ot_{1/2}$ value, reassociates **more** quickly than the *E. coli* genome, and is composed of **moderately** repeated sequences. Finally, the **slow** component reassociates with a $C_ot_{1/2}$ value more than 10^5 higher than the fast component. Like the *E. coli* genome, this component consists of **unique** sequence DNA. It reassociates much more slowly than *E. coli* DNA because there are many more different unique single strands that must find their correct opposite strands.

Expression of genome components

Which of these components contain genes that encode mRNA? This can be determined by including radioactively-labeled RNA as a **tracer** in the reassociation experiment. In this 'DNA-driven' experiment, **DNA** is present in vast excess to ensure that all the RNA will hybridize. When hybridization of the labeled RNA is measured, it is found to anneal to the **intermediate** and **slow** components. We therefore conclude that the **moderately** repeated and **unique** sequence DNA components are transcribed, with most of the transcription coming from the **unique** component.

Reassociation experiments can measure the expression of nonrepetitive DNA. To do so, the reassociation is performed in **RNA** excess (**RNA-driven**) and allows calculation of $R_ot_{1/2}$ values. These data are analyzed by plotting $R_ot_{1/2}$ value versus the proportion of DNA hybridized. Note that the theoretical maximum amount of DNA that can hybridize is **50%** of the starting DNA, as only one strand of the DNA is **complementary** to the RNA (the other strand of the DNA is **identical** to the RNA). Generally, these experiments show that a **small** amount of the DNA is able to hybridize to the RNA, indicating that a **small** portion of the unique sequence DNA is transcribed.

To study the complexity of the RNA population, RNA can be hybridized to **cDNA**. In this case, **all** of the **cDNA** will hybridize to the RNA. $R_ot_{1/2}$ is determined not by the number of copies of the genes, but by the number of copies of each RNA present in the RNA sample. As in $C_ot_{1/2}$ analysis, this $R_ot_{1/2}$ analysis detects **several** sequence components among the RNA population. In RNA isolated from the chicken oviduct, a large fraction of the cDNA hybridizes rapidly, indicating that the sample contains **many** copies of an individual RNA and its cDNA. This is the ovalbumin RNA, which is expressed at very **high** levels from a single copy gene. The next component represents seven or eight RNAs that are expressed at **moderate** levels. Finally, the high $C_ot_{1/2}$ component represents many other mRNAs that are expressed at **low** levels.

Reassociation experiments also demonstrate that while all cells have the same set of genes, **different** genes are expressed in different cells. While some genes are expressed in all cells (these are called **housekeeping** genes), others are expressed only in specific cells.

Chapter 22: Exons and introns

Answers to multiple-choice questions

1 (a,c); 2 (a,d); 3 (a,b,d); 4 (d); 5 (a); 6 (a); 7 (a,c,d).

Answers to concept questions

Question 1
The bulk of a typical mammalian gene consists of intron sequences; these are removed from the mRNA. In contrast, few yeast genes contain introns. Thus, most yeast mRNAs are not shortened by splicing and are roughly the same size as their genes.

Question 2
Different proteins can be generated from one gene by utilizing alternative promoters, alternative splicing, or alternative transcription termination.

Question 3

Selective pressure acts on exons because mutations that abolish function are selected against. Mutations in introns usually cause no phenotypes and are therefore not selected against, and are maintained in the gene pool.

Question 4

Zoo blots reveal how widely a DNA sequence has been conserved through evolution. Conservation generally indicates an important or functional DNA sequence.

Question 5

The functional domains of immunoglobulin proteins (discussed in Chapter 33) correspond well with the exons of the genes that encode these proteins. However, for other genes, exons do not correspond with functional domains of the encoded protein. Exon shuffling proposes that exons encode functional protein domains and can be rearranged by DNA recombination to create new genes. It would be much easier to build new genes by recombining existing domains, such as a membrane-spanning domain or an ATP-binding domain, than to evolve entire new proteins *de novo*.

Answers to problems

Problem 2

Refer to Figure 22.14.

Problem 3

Divide the 140 kb interval into smaller fragments and use each one to probe a zoo blot that contains DNA from other mammals. Fragments that are conserved between species will give a signal on the zoo blot; these are likely to encode exons. Use the exon fragments to probe a Southern blot containing DNA from patients with the disease. Often, disease results from DNA rearrangements within important genes. If patient DNA gives a different hybridization pattern from control DNA with one of the fragments, that fragment is strongly associated with the disease. Use that fragment to probe a Northern blot containing RNA isolated from patients and normal controls. If the amounts of RNA in patients and controls are different, this fragment probably contains the gene that is involved in the disease.

Answers to summary worksheets

Prokaryotic genes are colinear with their mRNA products. In striking contrast, eukaryotic genes are much **larger** than their mRNAs. This discrepancy is due to the presence of **introns** in most eukaryotic genes. Introns are removed by the process of **mRNA splicing**, which takes place after **transcription**. Although the gene and the mRNA differ in size, the **order** of exons within the mRNA does not change.

Introns are absent from **prokaryotic** genomes, relatively rare in **lower** eukaryotic genomes, and abundant in **higher** eukaryotic genomes. Higher eukaryotic exons tend to be **short** in length, while introns tend to be **longer**. Intron length also shows greater **variability** than exon length. The DNA sequence of exons is much **more** conserved than that of intron sequences.

How did introns arise? One model proposes that introns are ancient and have been lost from **prokaryotic** organisms, for whom rapid DNA replication is an advantage. Introns are maintained in higher eukaryotes because they allow generation of multiple proteins from a single gene through the process of **alternative splicing**. Introns have also been proposed to allow the facile generation of new genes through DNA recombination between introns in a process known as **exon shuffling**.

Chapter 23: Gene numbers

Answers to multiple-choice questions

1 (a,b,d); 2 (a,b,c,e); 3 (a); 4 (b); 5 (b,c,d,e); 6 (c); 7 (d,e,f); 8 (a); 9 (a); 10 (b,c).

Answers to concept questions

Question 1

Approximately 5000 darkly-staining bands can be visualized on *Drosophila* polytene chromosomes. In some cases, bands are thought to be sites of active transcription (discussed more thoroughly in Chapter 26). This number of bands correlates roughly with the estimated number of genes in the *Drosophila* genome, suggesting a one-to-one correspondence between bands and genes.

Question 2

M. Goebl and T. Petes (*Cell* 46, 983–992, 1986) inactivated yeast genes at random and determined the phenotype caused by gene inactivation (genes were inactivated by the insertion of foreign DNA). Only 12% of the insertions caused lethality, indicating that only this fraction of the yeast genome comprises genes essential for life. Another 14% of the genes were found to be important but not essential: inactivation of these caused the yeast to grow slowly. Surprisingly, 70% of yeast genes could be inactivated without noticeable effect on cell viability. These genes might perform a nonessential function or a function that is required only under certain conditions (for example, growth on a particular carbon source such as galactose). Some genes might perform important functions but are dispensable because other genes have the same function (redundant function) or similar functions (overlapping function).

Question 3

Adult mammalian hemoglobin is an $\alpha_2\beta_2$ tetramer. In α-thalassemia, too few α chains are produced and a β_4 tetramer forms. This tetramer does not function well.

Question 4

Globin genes arose through duplication of an ancestral gene, followed by transposition to a new locus and additional duplications. Pseudogenes arose when one copy became inactivated by mutation.

Question 5

Unequal crossing-over within the α or β locus can cause various deletions including a deletion between $\alpha1$ and $\alpha2$, deletion between $\psi\alpha$ and $\alpha2$, or between δ and β.

Question 6

Leghemoglobin is a single gene in plants. It differs from the single globin gene in primitive fishes by the presence of an intron. Duplication and diversion of the single globin gene of primitive fish led to the linked α and β genes of *Xenopus*. Further duplication generated two clusters of genes, and additional duplication and divergence led to the mammalian clusters.

Question 7

Processed pseudogenes have hallmarks of RNA processing reactions, indicating that at some point, they have gone through an RNA stage. For example, some processed pseudogenes lack introns, while others have been processed on their 3' ends. Presumably, they arose when a gene was transcribed into pre-mRNA, processed, and then copied back into DNA by reverse transcriptase. The reverse-transcribed copy then integrated back into the genome.

Question 8

The nontranscribed spacer of rRNA lies between the tandem transcription units, while the transcribed spacer is found between the 18S rRNA gene and the 28S rRNA gene of the transcription unit. Introns are embedded within the coding regions of genes. In contrast, transcribed spacers do not interrupt genes, but lie between the genes of a transcription unit. While introns are removed by splicing processes (Chapters 30 and 31), transcribed spacers are removed by a series of endonucleolytic cleavages (Chapter 12).

Question 9

It might be advantageous to transcribe 18S and 28S rRNAs together because they are needed in precisely the same amount.

Answers to problems

Problem 1

How many genes are expected to be found in the human genome (refer to Table 23.1)? How large is the average human gene? Calculate how much DNA is required to encode all these genes. How large is the human genome? Calculate the percentage of the human genome that encodes genes. How would this number be different if you calculated the percentage of the genome that encodes mature gene products (for example, mature mRNA or mature rRNA)?

Problem 2(a)

Hint: bands that hybridize strongly have DNA sequences very similar to the probe. Weak bands have less sequence similarity to the probe.

Problem 2(b)

Hint: here, each probe will hybridize strongly to its own mRNA if that mRNA is present.

Problem 3

To identify replacement and silent changes, refer to the genetic code in on inside back cover.

Problem 4

Figure 21.8 shows this experiment for mRNA. How would it differ if rRNA was used as a tracer? How highly repetitive are rRNA genes?

Answers to summary worksheet

The human genome contains approximately **700** times more DNA than the *E. coli* genome, yet is estimated to have 125,000 genes, only **50** times more than *E. coli*. Human genes tend to be **larger** than bacterial or yeast genes, primarily due to the presence of **introns**. Intriguingly, studies in yeast and *Drosophila* show that **few** genes are essential for the viability of the organism.

Genes in higher organisms **are not** all unique; many are present as members of gene **families**. When such genes are located adjacent to one another they form a gene **cluster**. Human globin genes are found in **two** clusters called the α-**cluster** and the β-**cluster**. Each contains multiple copies of **related** globin genes, which are expressed at different times during **development**. Clusters arose by the process of gene **duplication**. This occurs when chromosomes pair **improperly** prior to DNA recombination, in a process called **unequal** crossing over. This process generates two new chromosomes known as **nonreciprocal** recombinant chromosomes: one has lost a gene from its cluster while the other has gained a gene. DNA rearrangements that delete genes from the globin cluster can cause diseases known as **thalassemias**, in which insufficient globin proteins are synthesized.

During evolution, members of a gene family can change by the accumulation of **mutations**. When these occur at **replacement** sites, the amino acid sequence of the encoded protein is altered. When they occur at **silent** sites, no change in the encoded protein is caused. For active genes that are under selective pressure, changes will be found more frequently at **silent** sites. The degree of polymorphism between members of a gene family can be used to estimate the **time** since gene duplication.

Some members of gene families, called **pseudogenes**, are not expressed. Frequently, they contain mutations that **inactivate** them. Some pseudogenes, called **processed** pseudogenes, have passed through an **RNA** stage.

Chapter 24: Organelle genomes

Answers to multiple-choice questions

1 (a,b,c,d,e,f); 2 (c); 3 (b); 4 (a); 5 (a); 6 (a); 7 (a,c,d,e); 8 (a,c,d); 9 (a,c,d).

Answers to concept questions

Question 1
Mitochondria and chloroplasts contain their own genomes. Both these organelles have a unique internal environment that differs from the cytoplasm. Some proteins might be encoded in the organelle because they need this environment to fold correctly. Others may need to be produced inside the organelle because they would not be able to traverse the organelle membrane.

Question 2
In mammals, the mutation rate for mitochondrial DNA is higher than for nuclear DNA. However, in plants, mitochondrial DNA mutates at a lower rate then nuclear DNA. These differences probably arise because mitochondria use different DNA polymerase and DNA repair systems than the nucleus does.

Question 3
Economical features of the human mitochondrial genome include its very small size, genes that abut directly or even overlap, the presence of just one promoter region, and some genes that do not even contain full stop codons.

Question 4
Generally, only proteins are imported. However, no tRNA genes are evident in the mitochondrial genome of *Trypanosoma brucei*, suggesting that tRNAs might be imported into this organelle.

Question 5

The antibiotic sensitivity of organelle protein synthesis is similar to that of prokaryotes. In addition, organelle ribosomal proteins and RNA polymerase subunits are homologous to those of *E. coli.*

Question 6

rho⁻ yeast strains have large deletions and rearrangements of their mitochondrial genomes. The remaining DNA amplifies to high copy number.

Question 7

In mammals, all mitochondria are derived from the ovum.

Answers to problems

Problem 2

Indirect evidence for polyadenylated mRNAs in mitochondria comes from human mitochondrial genes that do not encode stop codons. These genes end with U or UA. It is presumed that polyadenylation adds additional A residues to create UAA stop codons. To directly test mitochondrial mRNAs for poly(A) tails, they could be hybridized to a column containing oligo(dT). Only RNAs with regions of poly(A) could bind to this column.

Problem 3

A promoter in the D-loop region generates a large counterclockwise transcript, which is processed to generate ND6 mRNA and several tRNAs.

Problem 4

Nuclear pre-mRNA splicing requires many *trans*-acting proteins and RNAs that are not encoded in organelle genomes. Do you think all these splicing factors could be imported? What *trans*-acting factors are required by self-splicing introns?

Answers to summary worksheet

Mendel discovered that genes segregate **independently**, with each parental allele inherited by **50%** of the offspring. Segregation of Mendelian traits is governed by the mitotic spindle, which ensures that each daughter nucleus receives one set of chromosomes from each parent. In non-Mendelian inheritance, parental contribution is **unequal**. The genetic information governing non-Mendelian traits is inherited from the **cytoplasm** instead of the **nucleus**.

Two organelles subject to non-Mendelian inheritance are the **mitochondria** and the **chloroplast**. Both function in **energy** conversion, are surrounded by unique **membranes** and carry their own **genomes**. Sequencing of organelle genomes reveals the presence of **a few** genes, primarily structural **RNAs** involved in **translation** and a few **proteins**. Most proteins found in the mitochondria are encoded by **nuclear** genes and must be **imported** into the organelle. The yeast mitochondrial cytochrome *c* oxidase is an enzyme that contains some subunits encoded by **nuclear** genes and some encoded by **mitochondrial** genes. It is interesting to speculate how nuclear and mitochondrial genes have evolved together such that their gene products are able to interact to form a functional enzyme.

Studies with inhibitors show that translation inside the organelles is more similar to **prokaryotic** than **eukaryotic**. This observation suggests that the mitochondria and chloroplasts in eukaryotic cells are **endosymbionts**. They were prokaryotes that were captured and maintained; some of their genes have been transferred to the nucleus.

Chloroplast genomes generally encode more gene products than **mitochondrial** genomes, including some subunits of **RNA** polymerase. The chloroplast genome is organized into two **single copy** sequences separated by an extended **inverted repeat**.

The yeast mitochondrial genome is **larger** than the human one. Many yeast mitochondrial genes contain **introns**; some of these contain **open reading frames**. In contrast, mammalian mitochondrial genomes are compact and contain no introns. Strikingly, in many cases there is no **separation** between genes. In fact, some genes even **overlap** by one or more nucleotides, reflecting a remarkable economy of genome use.

All but one human mitochondrial gene are **transcribed** in the same **direction** from a single promoter. The RNAs are derived from a single precursor transcript by **processing**. The genome is organized such that **protein-encoding** genes usually alternate with **tRNA-encoding** genes. Processing of the **tRNAs** generates individual mRNAs for the protein-encoding genes.

In yeast, the loss of mitochondrial function is not **lethal**, but does cause a phenotype called **petite**. Without mitochondria, the cells grow **anaerobically**, and their growth rate is lower. Nuclear petites lose mitochondrial function by mutation of a gene encoding an **imported** protein. Mitochondrial petites of the *rho⁰* type have lost all mitochondrial DNA while *rho⁻* petites have suffered a large **deletion** of mitochondrial DNA followed by **amplification** of the remaining sequences.

Chapter 25: Simple sequence DNA

Answers to multiple-choice questions

1 (c,d); 2 (a,c); 3 (a); 4 (b,c,d,e); 5 (b); 6 (a,b,c).

Answers to concept questions

Question 1

In density gradient centrifugation, each DNA fragment 'floats' to the part of the gradient where it has a neutral buoyany density. Each DNA fragment has a particular density based on its sequence. The main band of DNA is broad, reflecting many different sequences that differ slightly in their densities. If one sequence is present in very large amounts, it will form its own band.

Question 2

Digestion of mammalian DNA with a restriction enzyme produces a set of restriction fragments of many different sizes. These fragments produce a long smear on a gel. However, if a repeated sequence contains a restriction site, many fragments of identical size are generated. These form discrete bands visible within the smear.

Question 3

The mouse satellite repeat unit is 234 base pairs in length. It contains two closely related half-repeats of 117 base pairs, each of which is composed of two related 58 base pair sequences that are termed quarter repeats. The quarter repeats themselves are made of repeated sequences (one-eighth repeats), which are themselves composed of three related copies of a nine base pair sequence. These repeated sequences arose by duplication, amplification and divergence of an ancestral sequence.

Question 4

Saltatory replication generates tandem copies of a DNA sequence. For example, saltatory replication of the mouse 27 base pair repeat generated the 54 base pair repeat, which is composed of two tandem copies of the 27 base pair repeat.

Question 5

The sequence homogeneity of satellite DNA is maintained by crossover fixation, in which frequent unequal crossing-over leads the spread of one repeat unit and the elimination of others.

Answers to problems

Problem 1

Why do you think the major and minor satellite bands migrate so differently during equilibrium density centrifugation?

Problem 2

Hints: the minisatellite loci near the disease-associated gene are probably different sizes in each parent. How could you detect the sizes of these DNA fragments in (a) the parents; (b) the children?

Answers to summary worksheet

Satellite DNA is identified by the technique of **density gradient centrifugation**. It differs from the bulk of the DNA due to its **GC content**. Using *in situ* hybridization, satellite DNA is mapped to regions surrounding the **centromere**. This region of the chromosome is constitutive **heterochromatin**, and **is not** expressed.

Satellite DNA is predominantly composed of a **simple** sequence repeated many times. Mammalian satellite sequences are more complex than those of lower species and contain a **hierarchy** of repeating units. Despite considerable variation, the basic **nine** base pair repeat can be fitted into a **consensus** sequence. Satellite sequences have evolved by successive lateral or **saltatory replication** of the nine base pair repeat to form larger repeats. Variation is introduced into the repeats by **mutation**.

Repeated sequences are prone to **misalign** when chromosomes pair. This can lead to **unequal crossing-over**, in which the number of repeats on one chromosome **increases** while the number of repeats on the other chromosome **decreases**. The **crossover fixation** theory proposes that identical repeats are created and maintained by **unequal crossing-over** until one repeat becomes **dominant**.

Minisatellite sequences, like satellite sequences, are composed of **tandem** repeats. However, minisatellite sequences differ in that they contain **fewer** repeats, generally **five to ten** copies. The consensus core repeat is **ten to fifteen** nucleotides long and is rich in **GC** base pairs. The number of repeats found at a particular minisatellite locus is highly **variable** in human populations, making them very useful for **genetic mapping**.

Chapter 26: Chromosomes

Answers to multiple-choice questions

1 (c); 2 (a,c); 3 (b); 4 (b,c,d); 5 (a,b,d); 6 (a); 7 (c); 8 (b,c).

Answers to concept questions

Question 1

Packing ratio is the ratio of the length of a linear piece of DNA to the length of the same DNA in its compacted form.

Question 2

The first step in phage DNA insertion is translocation, in which the DNA is inserted into the phage head. A second condensation step follows, in which the DNA is compacted. Both processes require ATP.

Question 3
Prokaryotic DNA is organized into DNA–protein complexes called nucleoids. Proteins bind the DNA and organize it into large loops.

Question 4
HU binds DNA and is thought to wrap it into beads. In this way, it could be similar to a primitive histone protein.

Question 5
Yes, specific DNA sequences called matrix attachment regions bind to proteins of the nuclear matrix. These DNA sequences do not have a strong consensus sequence, although they are generally AT-rich.

Question 6
A chromosome scaffold is a dense, fibrous 'skeleton' of a chromosome. Scaffolds are formed when proteins are extracted from metaphase chromosomes. Large loops of DNA remain attached to the scaffold. The nuclear matrix is a fibrous network attached to the inside of the nuclear envelope. It is formed by extraction of proteins from interphase nuclei. Thus, the scaffold is present in mitotic cells, and the matrix is present in interphase cells. These structures share some common proteins, but each also contains proteins not found in the other.

Question 7
Kinetochore and centromere both refer to the point at which a chromosome attaches to the mitotic spindle. A kinetochore is defined by microscopy as a dense region of a chromosome where microtubules attach. A centromere is defined as a region of the chromosome required for proper segregation.

Question 8
Topoisomerase II is present in both these structures, but its precise role is unknown.

Question 9
The *CEN* sequence brings a functional centromere to the plasmid, allowing it to attach to the mitotic spindle and to be faithfully inherited as if it were a natural chromosome.

Question 10
Mutations in CBE-II are tolerated but mutations in CBE-III are not. *CEN* sequences bind a protein complex known as Cbf-III, which has the ability to translocate along microtubules. Perhaps this protein complex is responsible for moving daughter chromosomes along the mitotic spindle.

Question 11
Replication can proceed to the very 3' end of the chromosome by DNA polymerase in the 5' to 3' direction. However, DNA polymerase cannot replicate the 5' ends of chromosomes because there is no place for a primer to anneal. Thus, the DNA at the very 5' ends would be lost. This is

one reason why telomeres are added; telomeric DNA can be replaced by telomerase, which does not require an external template.

Answers to problems

Problem 1

The mutant phage would be replication deficient. Proheads would form but would never be filled with DNA. This could be detected by electron microscopy.

Problem 3

Refer to Figure 26.7. After isolating DNA associated with the nuclear matrix, prove that these sequences are sufficient for matrix association by adding them back to DNAase-treated matrix and demonstrating that they can still bind the matrix.

Problem 4

Note that yeast telomerase RNA would still be present, so a mixture of yeast and *Tetrahymena* telomeric repeats would be found. This experiment requires that the *Tetrahymena* telomerase RNA could associate with yeast telomerase proteins to produce a functional hybrid enzyme.

Answers to summary worksheet

All living cells must **condense** their genetic material. Viruses insert their genetic material into a **capsid**. Bacterial cells organize their DNA into a dense region known as a **nucleoid**. The DNA is held in large loops that define independent **domains**. The DNA is bound by proteins such as **HU**, which are thought to condense the DNA into bead-like structures.

Eukaryotic genomes are organized into **chromosomes**, which condense the DNA and ensure their faithful partitioning to daughter nuclei during **mitosis**. In interphase, chromosomes are attached to a fibrous network on the inner nuclear membrane called the **nuclear** matrix. Mitotic chromosomes consist of a dense network of fibers that anchors loops of DNA. This network is known as a **scaffold**.

Eukaryotic chromosomes are most readily visualized during **mitosis**, when they are **5–10** times more compact than in interphase. **Heterochromatin** stays condensed throughout the cell cycle while **euchromatin** becomes much less condensed. This is thought to facilitate **gene expression**. **Constitutive** heterochromatin is never expressed while **facultative** heterochromatin is selectively inactivated.

Polytene chromosomes of *Drosophila melanogaster* form when a synapsed pair of chromosome undergo **DNA replication** but do not **separate**. They show a distinctive pattern of **bands**. Some bands transiently expand or **puff**, due to accumulation of **proteins**. Puffs are thought to be sites of active **transcription**, and are therefore thought to represent active genes. Individual genes can be mapped to specific bands by the process of *in situ* **hybridization**.

Every chromosome needs three functional regions: a **centromere** for attachment to the mitotic spindle, an **origin** of DNA replication and **telomeres** to prevent loss of genetic information from the chromosome ends. Centromeres contain the **kinetochore,** to which the microtubules attach during mitosis.

Telomeres have a simple repeated DNA sequence that is rich in C and A residues on the 5' end. Different species have **different** repeat patterns. **Telomerase,** the enzyme that synthesizes telomeres, is unusual in that one component is composed of **RNA.** This component of telomerase provides the **template** for telomere synthesis. The template is complementary to **1.5** telomeric repeats. After one full repeat is added, the enzyme **shifts** forward to add a new repeat. The **3'** end of the G+T strand protrudes 14–16 nucleotides past the 5' end of the C+A strand. In this single-stranded region, G residues form an unusual hairpin structure known as a **G quartet.**

Chapter 27: Nucleosomes

Answers to multiple-choice questions

1 (a); 2 (c); 3 (e); 4 (f); 5 (a); 6 (b); 7 (e); 8 (b,c); 9 (a,c,d); 10 (a); 11 (a,b,d,e,f); 12 (a,c); 13 (a,b,c,d).

Answers to concept questions

Question 1

The nucleosome contains a histone octamer (two molecules each of H2A, H2B, H3 and H4) and 200 nucleotides of DNA. The 10 nm fiber is a longer piece of DNA with nucleosomes positioned every 200 nucleotides. The 30 nm fiber is composed of a 10 nm fiber and histone H1.

Question 2

Your diagram should look like Figure 27.22 except that the path of DNA around the nucleosome is not smooth.

Question 3

High conservation of amino acid sequence suggests that the protein performs an essential function and that most mutations are not tolerated.

Question 4

To find the amount of DNA tightly associated with a core particle, digest chromatin with increasing amounts of micrococcal nuclease and analyze the digestion products on a gel. At low concentrations, you will see a ladder that represents cutting between nucleosomes. As the nuclease concentration increases, the bands in the ladder will 'chase' into a single band; this is the DNA present in a trimmed nucleosome. Further digestion will result in a small decrease in the size of the band. This is the DNA in the core particle (approximately 146 nucleotides) that is tightly associated with the core particle.

Question 5

The H3$_2$•H4$_2$ kernel lies across the equatorial plane of the octamer, with one H2A•H2B dimer above the kernel and one below it.

Question 6

H1 differs from the core histones in several ways. It is not part of the core octamer; rather, it is found outside the octamer associated with the DNA as it enters and leaves the octamer. H1 is not as highly conserved as the other histones, and is modified at different times during the cell cycle from the core histones. H1 is required to form the 30 nm fiber.

Question 7

Cells were grown in medium containing heavy isotopes of carbon and nitrogen and switched to medium with light isotopes prior to DNA replication. Chromatin was isolated from the cells and the histone octamers were crosslinked (why was the crosslinking step essential for this experiment?). The density of the crosslinked histone octamers was determined by centrifugation. The crosslinked octamers were found to have intermediate density, indicating that the 'old' octamers had come apart and reformed randomly with 'new' histones. Electron micrographs of replicating chromatin reveal nucleosomes on newly replicated DNA, suggesting that nucleosomes form immediately after replication.

Question 8

Acetylation and methylation of lysine residues remove a positively charged amino group. Phosphorylation of the hydroxyl group of serine residues adds a negatively charged group. All these modifications reduce the positive charge of the histone protein, making it less able to cancel out the negative charge of the DNA backbone.

Question 9

The majority of H1 phosphorylation occurs before mitosis and the phosphates are removed when cell division is complete. During mitosis, chromosomes condense to a packing ratio of 10,000. It is tempting to speculate that H1 phosphorylation is involved in this condensation.

Question 10

Some nucleosomes are positioned on specific sequences. For example, in Figure 27.37 we see that nucleosomes are positioned specifically on the yeast URA3 gene when it is inactive. This positioning is lost when the gene is expressed. Positioning can result from an intrinsic feature of the DNA that favors nucleosome formation. For example, DNA sequences can have intrinsic bends that facilitate nucleosome formation. Alternatively, nucleosomes could be positioned by neighboring nucleosomes.

Question 11

Refer to Figure 27.42, in which the chromatin of adult and embryonic β-globin genes is probed in adult cells. The adult β-globin gene is very sensitive to digestion by low concentrations of DNAase I (the adult β-globin band is greatly diminished by treatment with 0.05 mg/ml DNAase I). In contrast, the embryonic β-globin gene, which is inactive in adult cells, is not digested until the DNAase I concentration reaches 0.5 mg/ml. Thus, the active gene is hypersensitive to DNAase I.

Answers to problems

Problem 1 (b)

Lane	DNase concentration	Marker sizes
1.	10.0 mg/ml	
2.	0 mg/ml	800 nt
3.	1.0 mg/ml	
4.	0.1 mg/ml	400 nt
		200 nt
		50 nt

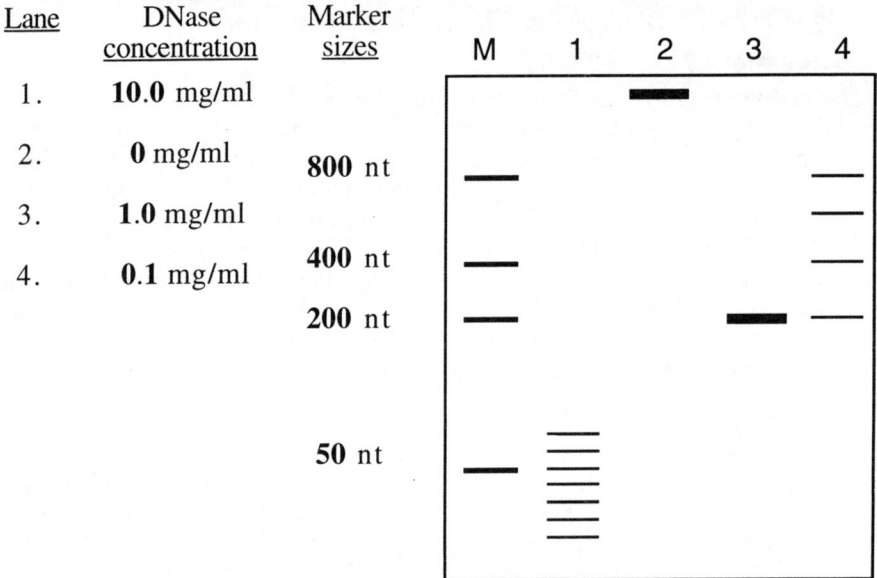

Problem (1c)

When no DNAase I is added, the chromatin is not digested and a high molecular weight band is seen. At low concentrations (0.1 mg/ml), DNAase I cuts in the linker DNA between nucleosomes to generate a ladder of mononucleosomes (200 nt band), dinucleosomes (400 nt band), etc. Increasing the nuclease concentration to 1.0 mg/ml results in cutting all the multi-nucleosome bands into mononucleosomes. At very high DNAase I concentrations, the DNA that is wrapped around the histone octamer can be cut. Here, each strand is susceptible only when it is on the outside of the double helix. This generates a ladder of bands separated by 10 nucleotides, corresponding to the structural periodicity of DNA.

Problem 3(a)

The chromatin structure of the *ACT1* gene is very sensitive to nuclease digestion, as is expected for actively transcribed genes. *ACT2* is transcribed at a lower level and is partially resistant to nuclease. The pseudogene *ACT3* is not expressed and is not degraded by DNAase I, indicating that *ACT3* lies in chromatin structure that is inaccessible.

Problem 3(b)

When growth conditions are changed, both *ACT1* and *ACT2* have intermediate DNAase I sensitivity. This finding predicts that the *ACT1* and *ACT2* mRNAs will be found in comparable amounts.

Answers to summary worksheet

Nucleosome structure

Nucleosomes are the basic structural unit of chromatin. They consist of a **histone** core and a length of DNA. The core contains eight **histones**: two molecules each of **H2A, H2B, H3** and **H4**. Histones are **positively** charged proteins that are highly **conserved** among all eukaryotes. Histones undergo transient post-translational **modification** including **acetylation, methylation** and **phosphorylation**. All these modifications make the charge of the histones more **negative**, which is thought to alter their interaction with **DNA**.

Nucleosomes in chromatin are spaced about **200** nucleotides apart. This is determined by digesting chromatin with a low concentration of **micrococcal nuclease**. If a higher concentration of enzyme is used, the chromatin is digested to a length of **165** nucleotides, which represents the **trimmed** nucleosome. Further digestion reduces the DNA to **146** nucleotides. This is the **core** DNA, which is protected from digestion by the nucleosome. In contrast, the **linker** DNA between nucleosomes is readily digested. When DNA is **wrapped** around a histone core, DNA sequences **80** nucleotides apart are close together.

Digestion with high concentrations of nuclease changes the cleavage pattern: now cleavage is observed every **10** nucleotides. This cleavage periodicity corresponds to the **structural** periodicity of the double helix, demonstrating that the DNA is wrapped around the **outside** of the histone core.

Isolated histone octamers have **the same** shape as from core nucleosomes (which contain a histone octamer and **DNA**). This indicates that nucleosome structure is largely dependent on the **histone octamer**. The structure of the histone octamer has been studied by **X-ray crystallography**. It consists of an **H3$_2$•H4$_2$** tetramer and two **H2A•H2B** dimers, one **above** the tetramer and the other **below** it. The path of the DNA in the nucleosome **is not** smooth.

Gentle extraction of chromatin under low **ionic strength** (low salt concentration) yields the **10** nm fiber. In this form, chromatin resembles

273

beads on a **string**. At higher ionic strength, the **30** nm fiber is seen. This requires the presence of **H1**, and is a **solenoid** of the **10** nm fiber, with **six** nucleosomes per turn. The 30 nm fiber has a **packing** ratio of **40**. Histone H1 **is not** part of the core particle, and is thought to **seal** the DNA in the core particle.

Nucleosome formation

Upon DNA replication, nucleosomes reform **immediately**. The 'old' histones, which were on the DNA before replication, **dissociate** and **reform** into 'new' nucleosomes. This was determined by growing cells in medium containing **heavy** isotopes of carbon and nitrogen, then switching to medium with **light** isotopes prior to DNA **replication**. Chromatin was isolated from the cells and the histone octamers **crosslinked** to preserve their associations. The **density** of the crosslinked histone octamers was determined by **centrifugation**. If the old octamers remained **associated**, a pool of **heavy** 'old' octamers would be found at the **bottom** of the gradient and a pool of **light** 'new' octamers at the **top**. If the octamers **dissociate** and **reform**, crosslinked octamers of **intermediate** density would be found. Nucleosomes can form by two pathways: an **H3•H4** kernel can form, bind **DNA** and finally bind **H2A•H2B**. Alternatively, the entire **octamer** can preform prior to DNA binding.

Nucleosome positioning

A critical question is whether nucleosomes are found at specific DNA sequences, and how this could affect gene **expression**. This question of nucleosome **positioning** has been addressed using the technique of **indirect end-labeling**. In this procedure, chromatin is first treated with **micrococcal nuclease**, which cuts **between** nucleosomes. The histones are then removed and the DNA digested with a **restriction enzyme**. DNA fragments are separated by gel electrophoresis and the fragment of interest detected by **Southern** blotting. The fragment of interest will have one end generated by **micrococcal nuclease** and the other end by the **restriction enzyme**. If the nucleosomes are positioned **specifically**, the **micrococcal nuclease** will always cut the same distance from the **restriction enzyme** and the fragment of interest will appear as a **discrete** band. If nucleosomes are positioned randomly, the **micrococcal nuclease** cuts will be various distances from the **restriction enzyme** site, and the fragment of interest will appear as a **diffuse** band. These experiments always require a control in which **naked** DNA is treated in parallel. This control detects DNA sequences that the nuclease cleaves **preferentially**.

Nucleosome positioning can be an **intrinsic** property of the DNA, as some DNA sequences favor nucleosome formation. Alternatively, nucleosome positioning can be an **extrinsic** property of the DNA, where nucleosome position is based solely on the distance from other nucleosomes.

Nucleosomes and transcription

Another important question is whether DNA covered by nucleosomes can be **transcribed**. RNA polymerase must **unwind** the strands of the DNA double helix before it can make RNA; it is not clear how DNA in a nucleosome could be unwound. Experiments with micrococcal nuclease show that actively transcribed genes have **the same** nucleosome density as latent genes. Additional experiments suggest that nucleosomes are **displaced** by RNA polymerase, and **reform** behind the enzyme after it passes. RNA polymerase moves quickly through the first **30** nucleotides of a nucleosome, then begins to **pause** after the addition of each ribonucleotide. When the enzyme reaches **halfway** through the nucleosome, it speeds up again. This could be the point at which the nucleosome is **transferred** to a new position behind RNA polymerase. Transcription **can** proceed through a crosslinked octamer, indicating that the octamer **does not** need to dissociate in order to be **transferred**.

Very light digestion of chromatin with DNAase I reveals **hypersensitive sites**. These regions are thought to be especially sensitive to digestion because they have **no** nucleosomes. Hypersensitive sites are often found at **promoters** of genes, suggesting that nucleosomes are removed to allow transcription factors access to the DNA. A promoter is hypersensitive to DNAase I only in the **tissues** in which the gene is **expressed**.

Although active genes are covered by nucleosomes, they are **more** sensitive to nuclease degradation than inactive genes. For example, in adult chicken erythrocytes, the adult β-globin gene is degraded at **low** concentrations of DNAase I, while the embryonic β-globin gene is digested only at **higher** concentrations. In some cases, the hypersensitive region extends over more than just the gene. This suggests that chromatin structure is altered in **domains** containing active genes.

Chapter 28: Initiation of transcription

Answers to multiple-choice questions

1 (c,d,e); 2 (d); 3 (a); 4 (a,b,d,e); 5 (b,f); 6 (b,d,e); 7 (a,c,d); 8 (e); 9 (b).

Answers to concept questions

Question 1

TBP is a component of the polymerase I factor SL1, the polymerase II factor TFIID, and the polymerase III factor TFIIIB. Each of these proteins helps to position RNA polymerase near the initiation site. TBP might interact with a subunit that is shared among all three RNA polymerases.

Question 2

A preinitiation complex contains all the basal transcription factors required for RNA polymerase to bind.

Question 3

The factors TFIIIA and TFIIIC bind the internal polymerase III promoter and recruit TFIIIB to a site upstream of the initiation site. TFIIIB then positions RNA polymerase III at the initiation site.

Question 4

The internal A, B and C boxes must be recognizable by the polymerase III transcription factors that bind them. Thus, they cannot deviate very much from the consensus sequences. Because these elements are within the coding region of the polymerase III transcript, they constrain the coding capability of the transcript in that region.

Question 5

TFIIH is found in two forms: a kinase complex and a repair complex. After transcription initiation, the kinase form may be replaced by the repair form. RNA polymerase II stalls when it encounters damaged DNA. This may be the signal that activates the TFIIH repair activity.

Question 6

Introduce mutations into the TATAAAA sequence upstream of a reporter gene and test the transcription efficiency of each mutant *in vivo* or *in vitro*. Be sure to include an unmutagenized TATA box as a control.

Question 7

TFIIIA binds to the C box and recruits TFIIIC. TFIIIB is then positioned upstream of the initiation site. At this point, the preinitiation complex is stable and does not require the continued presence of TFIIIA or TFIIIC. This can be shown by carrying out the reaction *in vitro* using purified components. After TFIIIB is bound, TFIIIA and TFIIIC can be washed away using a high salt treatment. TFIIIB is still bound and able to recruit polymerase III.

Question 8

RNA polymerases I, II and III differ in subunit composition, sub-nuclear localization, type of genes transcribed and sensitivity to α-amanitin and other inhibitors. Some subunits are identical in all three polymerases while other subunits are homologous. All three polymerases interact with common proteins such as TBP. They differ from prokaryotic RNA polymerase by having many more subunits, although they contain subunits that are homologous to those of prokaryotic RNA polymerase.

Question 9

Immunoglobulin genes might require Oct-2 or other transcription factors that are present only in lymphoid cells. The presence of a transcription factor is not always sufficient to activate a gene: the entire context of the promoter must be considered. Also, as we will see in Chapter 33, immunoglobulin genes are not fully active until DNA rearrangement brings a downstream enhancer within working distance of the promoter.

Answers to problems

Problem 1

To map the SP1 sites, set up the following experiment:
- (i) Label the DNA fragment on its 5' ends with kinase or on its 3' ends with Klenow DNA polymerase.
- (ii) Digest with one restriction enzyme to cut off one labeled end.
- (iii) Add SP1 in various amounts.
- (iv) Add DNAase I (which has been previously titrated to achieve about one cut per DNA molecule).
- (v) Analyze the digestion products on a denaturing gel. Be sure to include a 'no SP1' control to show which sites are naturally hypersensitive to DNAase I.

The site bound at lowest SP1 concentration has the highest affinity.

Problem 2

How many copies of FE must bind in order to activate transcription? What spatial arrangement do they require to be active?

Problem 3

What kind of genes do not require upstream sequences for their expression? What would happen if you delete into the coding region?

Problem 4

What kind of elements affect transcription over a large distance? If two DNA sequences are separated by a long distance of linear DNA, does that mean they are always far apart?

Problem 5

On linear DNA, TFIIE, TFIIH and ATP are required to unwind the DNA in front of the polymerase II complex. These factors are not needed for transcription of supercoiled DNA because it is energetically favorable to unwind the negative supercoils.

Answers to summary worksheet

Prokaryotes have **one** RNA polymerase while eukaryotic cells have **three**, each with a distinct function. Polymerase I makes **rRNA** and is localized to a sub-nuclear structure called the **nucleolus**. Polymerase II synthesizes **mRNA** and is found in the **nucleoplasm**. Polymerase III is also found in the **nucleoplasm** and synthesizes **tRNA** and **5S rRNA**. All three polymerases are large protein **complexes** with **multiple** subunits. Several subunits are **shared** by all three polymerases; others are specific to a single polymerase. The largest subunits are homologous to **prokaryotic** RNA polymerase.

RNA polymerase I

Polymerase I promoters are unique in that they are all essentially **identical**. The promoter contains two regions: the **core promoter** and the **UCE**. Both elements have a similar **sequence**, which is rich in **G•C** pairs. The polymerase I transcription factor **UBF1** binds to both regions. Next, the factor **SL1** binds cooperatively. When both factors are bound to the promoter, this is called a **preinitiation complex**, which is ready for the addition of RNA polymerase. While UBF1 is a **single** polypeptide, SL1 contains **four** subunits, including a protein called **TBP**. This protein is also required by **RNA polymerase II** and **RNA polymerase III**. It does not bind directly to the **DNA**; rather, it is thought to function by **recruiting** RNA polymerase to the preinitiation complex.

RNA polymerase III

RNA polymerase III promoters are unusual in that they usually are located **within** the genes. Two types of internal polymerase III promoters are found. Type I promoters are generally found in 5S rRNA genes and contain two conserved sequences termed **box A** and **box C**. Transcription factor **TFIIIA** binds box C and recruits factor **TFIIIC**. The next factor to bind is **TFIIIB**, which contains **TBP** and other proteins. This factor binds near the **initiation** site and recruits **RNA polymerase III**. Once **TFIIIB** is bound, the preinitiation complex is **stable** and does not require the continued presence of **TFIIIA** or **TFIIIC**. Therefore, TFIIIA and TFIIIC are termed **assembly** factors while TFIIIB is called an **initiation** factor.

Another type of polymerase III promoter is the upstream promoter, found at some genes encoding **small nuclear RNAs**. These promoters may contain the elements PSE, OCT and a **TATA** box. While the **TATA** box is sufficient for transcription initiation, the efficiency of transcription is increased when the other elements are present. In these cases, factor **TFIIIB** may recognize the **TATA** box directly. TFIIIB resembles polymerase I factor **SL1** in three ways. First, it does not bind DNA itself; second, it contains **TBP**; and third, it is thought to interact with RNA polymerase.

278

RNA polymerase II

RNA polymerase II promoters also consist of several sequence elements, although they are much more diverse than polymerase I or polymerase III promoters. First, a loosely conserved **Inr** (for **initiation region**) sequence is found near the transcription start site. Most polymerase II promoters also contain a **TATA** box approximately 25 nucleotides upstream of the start site. This element has the consensus sequence **TATAAAA** and, except for its location, is similar to the prokaryotic **−10** element. This is the binding site for **TFIID**, the first factor to recognize polymerase II promoters. This factor is a large complex composed of **TBP** along with roughly seven other subunits called **TAFs** (which stands for **TBP-associated factors**). Different combinations of **TAFs** create different forms of **TFIID** that might have different specificities. After TFIID has bound to the TATA box, **TFIIA** binds upstream of the TATA box, followed by **TFIIB**. This forms the polymerase II **preinitiation** complex, which now can bind RNA polymerase II and is competent to initiate transcription. However, additional factors **TFIIE** and **TFIIH** are required for the release of polymerase II from the promoter. **TFIIH** contains several enzymatic activities including a protein **kinase** activity. It can phosphorylate the **CTD** (which stands for **C-terminal domain**) of polymerase II. This phosphorylation is thought to trigger the switch from transcription initiation to transcription **elongation**. A minority of polymerase II promoters lack a **TATA** box. **TFIID** still binds these promoters, presumably due to association of one of the **TAFs** with the Inr.

The Inr and the TATA box constitute the **basal** polymerase II promoter. In addition to these DNA sequences, many other short DNA sequence elements are usually located **upstream** of the basal promoter and are recognized by specific transcription factors. In contrast to the basal elements, which define the **location** of transcription initiation, these upstream elements affect the **frequency** of initiation. Transcription factors bound to upstream elements function by **interacting** with basal factors. Some genes contain additional elements called **enhancers** that are located far from the basal promoter and contain a high concentration of **transcription factor** binding sites. They can be found **upstream** or **downstream** of the promoter, and can even be in the **opposite** orientation. They are thought to interact with the basal promoter when the intervening DNA forms a large **loop**. The net effect is to increase the **concentration** of transcription factors at the promoter.

Transcription factors commonly contain two independent **domains**: one for DNA **binding** and one for transcription **activation**. The activation domains can directly contact basal factors including **TAFs** or **TFIIB**. Alternatively, they may act indirectly through proteins known as **coactivators**.

279

Chapter 29: Regulation of transcription

Answers to multiple-choice questions

1 (c); 2 (a,b,c,d); 3 (c,e); 4 (a,c,e); 5 (a,b,c,e); 6 (d); 7 (f); 8 (b,c,e); 9 (b,c,d); 10 (b) (although Hpa II will not cleave if the site is methylated).

Answers to concept questions

Question 1

Activation of gene structure, 3' end formation, processing of the transcript, mRNA transport to the cytoplasm

Question 2

Steroid receptors bind zinc through a Cys-Cys-Cys-Cys motif while zinc finger proteins use a Cys-Cys-His-His motif.

Question 3

Leucine zipper transcription factors recognize inverted repeats with no separation between the half sites. The leucine zipper regions hold the two subunits together, positioning the adjacent basic regions next to each other and in the opposite polarity.

Question 4

Both homeodomain and leucine zipper transcription factors bind DNA through an α-helix positioned in the major groove.

Question 5

Genes under coordinate control all have a common response element in their promoters. When the appropriate transcription factor is activated, all genes with this response element are activated.

Question 6

Synthesis of the protein (homeodomain), phosphorylation (HSTF), dephosphorylation (AP-1), ligand binding/nuclear localization (steroid receptor), release of active factor (sterol), release from inhibitor (NF-κB).

Question 7

The conserved amino acids form the general structure of the DNA-binding domain, while the sequence-specific contacts are made by amino acids that vary in each family member, allowing each to recognize a different DNA sequence.

Question 8

Mutant transcription factors might inappropriately activate genes that stimulate cell division.

Question 9

Some can and some cannot bind to nucleosome-covered DNA. TFIIIA cannot bind DNA covered by nucleosomes, while the glucocorticoid hormone receptor can.

Question 10

In the preemptive model, transcription factors and nucleosomes cannot displace one another, except at DNA replication. At that time, transcription factors can stably bind to the DNA and allow transcription. In the dynamic model, histones are displaced by an ATP-dependent factor, allowing transcription factors to bind the DNA.

Question 11

The *SWI* and *SNF* gene products from a large complex that 'remodels' chromatin by moving or removing nucleosomes, thereby allowing transcription factors to activate many genes.

Question 12

Domains are bordered by insulators, which isolate the domain from other domains. Domains can also contain a MAR (matrix attachment region), which attaches the domain to the nuclear matrix. Finally, domains can have a LCR (locus control region), which has DNAase I hypersensitive sites and acts as an enhancer for the entire domain.

Question 13

Some genes are imprinted, or methylated, during spermatogenesis or oogenesis. IGF-II is methylated in oocytes and therefore the maternally inherited allele is not expressed in offspring. The paternal allele is not methylated and is expressed. Other genes are methylated only during spermatogenesis; for these genes, the maternal allele is expressed.

Question 14

MyoD is a bHLH protein expressed only in myogenic cells. MyoD homodimerizes poorly but is active as a heterodimer with bHLH proteins E12 or E47. MyoD activity can be controlled by the HLH protein Id. Id forms heterodimers with E12 and E47. These do not bind DNA, and prevent MyoD activity by removing its required dimerization partners.

Answers to problems

Problem 1

Find a conserved DNA binding motif in this sequence. Which amino acids are required for zinc binding? Which contact the DNA?

Problem 2

(a) Isolate a restriction fragment containing the 5S gene. Label both 5' ends with polynucleotide kinase, cleave off one labeled end with a restriction enzyme. Allow purified TFIIIA to bind the DNA, then treat with DNAase I and analyze the cleavage products on a gel. Sites of protein binding are identified by a region of DNA that is protected from digestion. Be sure to include a no protein control and to titrate the DNAase I.

(b)   Which residues would abolish zinc binding if mutated?

(c)   The DNA-binding domain is not affected by deletion of these amino acids. What other functional domains do transcription factors have? With what other proteins might these domains interact?

(d)   TFIIIA regulates the accumulation of 5S rRNA in *Xenopus* oocytes because it binds both the 5S rRNA gene and the 5S rRNA itself. When little 5S rRNA is present, most of the TFIIIA is available to bind DNA and stimulate transcription of 5S genes. As the level of 5S rRNA increases, TFIIIA binds the RNA and is not available for transcription. This is an example of feedback regulation.

Problem 3

(b)   HNF3 binds both naked DNA and chromatin. Which substrate does it prefer?

(c)   Perform an indirect end-labeling experiment as follows.

   (i)   isolate chromatin and add purified HNF3;

   (ii) digest chromatin with DNAase I (light digestion, titrate enzyme);

   (iii) remove proteins;

   (iv) cut with restriction enzyme near the HNF3 sites;

   (v)   run on gel and blot;

   (vi) probe for albumin upstream region.

   If nucleosomes are phased, you will see specific bands. If they are not phased, you will see a smear.

(d)   Adding or deleting five nucleotides between the HNF3 and GATA4 sites rotates the GATA4 site by one half turn of the helix. Because the HNF3 site is in the middle of the nucleosome-wrapped DNA, the GATA4 sites must also be within the same nucleosome. Thus, rotation of the site would place it inside (facing the histones) and thereby unavailable for binding GATA4.

Problem 4

The A and B boxes bind TFIIIC, which recruits TFIIIB. Interaction of TBP (in TFIIIB) with the TATA box helps stabilize TFIIIB at the upstream region. The TATA box would not be needed if the A and B boxes matched the consensus well or were properly positioned.

Answers to summary worksheet

Genes that are coordinately regulated often have a conserved **response** element in their promoters. For example, heat shock gene promoters contain a DNA sequence known as the **HSE**. This element is bound by the transcription factor **HSTF**, which allows all heat shock genes to be induced together. Some promoters contain **many** different response elements so that transcription can be activated in response to a variety of stimuli.

Transcription factors

The transcription factors that bind these elements are classified into families based on their **DNA-binding** domains. Although each type of DNA-binding domain is distinct, they are responsible for two common features of transcription factors. First, transcription factors usually bind DNA through an α-helix domain, and second, transcription factors of most families act as **dimers**.

Zinc finger proteins bind zinc through a conserved amino acid sequence containing the residues **Cys-Cys-His-His**. Most transcription factors of this family have **multiple** zinc fingers, which contact the DNA through an α-helix formed by the C-terminal side of the finger. Steroid receptors bind zinc through **four** conserved **Cys** residues. DNA binding is carried out by the C-terminal part of the finger while the N-terminal region is involved in **dimer** formation. These factors usually contain three domains: **transcription** activation, **DNA** binding and **hormone** binding. In the case of the glucocorticoid receptor, hormone binding causes a change in the intracellular **location** of the protein.

The homeodomain proteins share a 60 amino acid domain that forms three α-helices: the **third** one contacts DNA in the **major** groove. Transcription factors of this family are important regulators of **development** in many organisms.

HLH proteins are not named for a DNA-binding domain. Rather, the helix-loop-helix domain is involved in **dimerization** of these transcription factors. This domain forms two **amphipathic** α-helices separated by a loop region. Many members of this family are called bHLH proteins because they contain a region of **basic** amino acids next to the HLH domain. This is the DNA-binding domain. bHLH proteins form several types of dimers including homodimers, heterodimers with other bHLH proteins, and heterodimers with HLH proteins. This last class **does not** bind DNA with high affinity.

The leucine zipper family is similar to the HLH family in several ways. The **leucine** residues form an amphipathic α-helix that is responsible for **dimer** formation. They also bind DNA through an adjacent **basic** region, and form both **homodimers** and **heterodimers** have different DNA binding properties.

Effects of chromatin structure

Gene expression is dependent upon the state of **chromatin**; most genes covered by nucleosomes **are not** expressed. Some transcription factors may function by **displacing** histones from promoter region. In contrast, the MMTV promoter must be in a **nucleosome** for the transcription factors HR, NF1 and OTF-1 to bind. Large-scale chromatin structures called **domains** can also affect gene expression. Domains are bordered by **insulator** regions and can contain MAR sequences to attach the domain to the **nuclear matrix**. Within a domain, the **LCR** region contains the promoter and enhancer functions.

Finally, gene expression is also affected when DNA is modified by **methylation**. Usually, the **C** residue of a **CpG** doublet is methylated. In general, methylated genes are **inactive**. Methylation is also responsible for the phenomenon of **imprinting**, in which alleles of certain genes from one parent are never expressed.

Chapter 30: Nuclear splicing

Answers to multiple-choice questions

1 (c,e); 2 (a,c,e); 3 (d); 4 (e); 5 (a,b,c); 6 (a,b,d); 7 (a,c,d,e); 8 (a,b,c); 9 (b,c,d,e); 10 (a,b,d,e); 11 (b,c,d); 12 (b,d).

Answers to concept questions

Question 1
The branch site is found within a loose consensus sequence. It is identified primarily by proximity to the polypyrimidine region and the 3' splice site. When a 'real' branch site is moved away from the polypyrimidine region and the 3' splice site, it is no longer recognized by the U2 snRNP. Instead, an A residue near the polypyrimidine region will be used as the branch site.

Question 2
See Figure 30.10. In the E complex, the U1 snRNP binds to the 5' splice site and U2AF binds to the polypyrimidine tract. The U2 snRNP binds to the branch site to form the A complex. This is followed by addition of the U4/U6–U5 trimer to form the B1 complex. snRNPs rearrange (U5 shifts from exon to intron, U6 replaces U1 at the 5' splice site) to form the B2 complex. Further snRNP rearrangements (U4 is released, U6/U2 forms catalytic center) form the C1 and C2 complexes, which are the active forms of the spliceosome.

Question 3
U1 interacts with the 5' splice site by RNA–RNA base pairing between the 5' end of the U1 RNA and the 5' splice site consensus sequence. This was demonstrated by making mutations in the 5' splice site, which disrupt base pairing with U1 and inhibit splicing. Next, compensatory mutations,

designed to restore base pairing to the mutant 5' splice site, were made in U1 RNA. The compensatory changes in U1 RNA restored splicing activity of the mutant 5' splice sites, indicating that these RNAs interact by base pairing.

Question 4

U1–5' splice site; U2–branch site; U4–U6, U2–U6; U6–5' splice site; U5–exons.

Question 5

RNA helicases are needed to melt the many RNA helices that form and break during spliceosome assembly. These are probably needed at steps in spliceosome assembly where snRNAs rearrange; for example, when the U4 and U6 snRNAs disassociate.

Question 6

The catalytic center is formed by the U2 and U6 RNAs and the intron. This forms in the C complex of spliceosome assembly. The secondary structure created by the U2–U6 pairing looks similar to domain 5 of Group II introns.

Question 7

Group II and pre-mRNA splicing use the identical reaction mechanism, have similar splice sites, and form a similar secondary structure in their catalytic sites (see answer to question 6). Pre-mRNA introns could have evolved from Group II introns by having the domains split off into separate genes: these could have become the snRNAs. They could then interact to recreate a Group II catalytic center from separate pieces. This may be advantageous to the organism because the sequence of each intron would no longer need to be maintained. Intron sequences could be free to evolve into new genes. Also, it would allow alternative splicing, which is not possible in Group II introns.

Question 8

Sex determination involves a cascade of alternative splicing events depicted in Figure 30.19. In females, the ratio of sets of autosomes to X chromosomes causes the *Sex-lethal* gene to be spliced in the female pattern and Sxl protein is produced. The Sxl protein causes *transformer* to be spliced in the female mode, allowing the Tra protein to be made. The Tra protein promotes female-specific splicing of *transformer-2*, resulting in production of the Tra2 protein. The Tra2 protein causes female-specific splicing of the *doublesex* gene to make the female-specific Dsx protein, which acts to suppress male genes and cause female development. In males, no Slx, Tra or Tra2 proteins are produced. Therefore, the male-specific Dsx protein is made, which causes male differentiation.

Question 9

RNA polymerase I terminates at a discrete site one kilobase downstream from the mature 3' end, which is generated by an endonucleolytic cleavage. RNA polymerase III terminates when it encounters a sequence of four U residues (more accurately, this should be referred to as a run of four A residues in the template strand). The run of U residues must be within a GC-rich region, and transcription usually terminates with the second U. The 3' end is not processed further (except in the case of tRNAs, which are processed by a special series of enzymes). RNA polymerase II proceeds long past the site where the mature transcript ends. The 3' end is formed by cleavage downstream of the AAUAAA sequence followed by addition of the poly(A) tail.

Question 10

The processing of histone mRNA 3' ends requires an internal hairpin and base pairing with the U7 snRNA. The role of each secondary structure (internal histone hairpin and intermolecular histone–U7 helix) can be tested by mutating nucleotides on one side of the helix. These should disrupt processing. Now the other side of the helix is mutated to restore base pairing, which should restore processing. If so, this demonstrates that the secondary structures, not the primary sequences, are important. Recognition of the cleavage site by U7 snRNA is reminiscent of interactions between the U1 snRNA and the 5' splice site, and between the U2 snRNA and the branch site region.

Question 11

It might be advantageous to put the SL RNA on to all mRNAs so that they all would have an efficient translation initiation sequence.

Answers to problems

Problem 1

The order of intron removal can be determined by probing a Northern blot with a series of probes specific for different introns and exons, as seen in Figure 30.5.

Problem 2

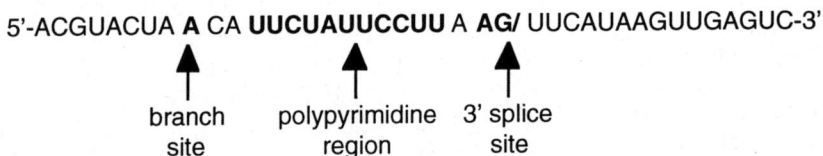

5'-ACGUACUA **A** CA **UUCUAUUCCUU** A **AG/** UUCAUAAGUUGAGUC-3'

branch   polypyrimidine   3' splice
site   region   site

5'-ACGUACUA A CA UUCUAUUCCUU A **CC** UUCAUA **AG/** UUGAGUC-3'

                     ↑   ↑

                  mutate   cryptic

                  normal   3' splice

                3' splice   site is

                   site   chosen

Problem 3

In males, where no Sxl protein is present, where does U2AF bind? Which splice site is then selected? In females, which protein will bind to the Py-NSS? Where will U2AF bind? Which splice site is selected now?

Problem 4

(a) U6 mutant 1 makes complex B1 but cannot go on because U4/U6 cannot come apart.

(b) U6 mutant 2 cannot make complex B2 because U6 cannot base pair with the 5' splice site.

(c) U6 mutant 3 can form complex C1 but cannot catalyze splicing.

(d) U6 mutant 4 can catalyze splicing but U6 cannot be recycled into U4/U6, resulting in depletion of the U6 available for splicing.

Problem 5

Of the nucleotides involved in the splicing reaction, which need to have a 2' OH group?

Answers to summary worksheet

Pre-mRNA splicing

Introns are bordered by consensus sequences: **GU** at the 5' splice site and **AG** at the 3' splice site. In addition, an essential **adenosine** nucleotide, called the **branch** site, is found within a loose consensus sequence near the 3' end of the intron. Most introns also have a region rich in **pyrimidine** residues located between the branch site and the 3' splice site.

In the first step of splicing, the **2' OH** group of the branch site adenosine attacks the **5'** splice site. This attack breaks the bond between the last nucleotide of the **exon** and the first nucleotide of the **intron**. Concomitantly, a new 2'–5' bond is formed between the first nucleotide of the **intron** and the **branch site** nucleotide, creating a **lariat** structure. The 5' exon now ends in a 3' **OH** group, which initiates the second step of splicing by attacking the **3'** splice site. This attack breaks the bond between the last nucleotide of the **intron** and the first nucleotide of the 3' **exon**, while simultaneously forming a new bond between the **exons**. The end products are **mature** mRNA and released intron, still in the **lariat** form.

The splice sites are recognized by **snRNP** particles, which contain both **RNA** and **protein** components. The snRNPs **assemble** onto the pre-mRNA in a stepwise pathway to form a large multi-component complex called the **spliceosome**, in which splicing occurs. First, the **U1** snRNP binds to the 5' splice site. Also at this stage, the protein splicing factor **U2AF** binds to the **polypyrimidine** tract near the 3' end of the intron. Next, the **U2** snRNP binds to the **branch site**. Recognition of the 5' splice site and the branch site relies in part on **RNA–RNA** interactions between the intron and the snRNAs. The next step is the addition of the **U4/U6-U5** trimer to the complex, followed by a series of **conformational** rearrangements. Important changes that occur during these steps include disruption of the **U4/U6** interaction and formation of a **U2/U6** pairing. In addition, **U1** is displaced from the 5' splice site by **U6**. These processes culminate in the creation of an RNA **catalytic** center, which can then initiate the first step of splicing.

The identical reaction mechanism is used by Group **II** introns, which are autocatalytic. Unlike pre-mRNA introns, Group II introns fold into an elaborate **secondary** structure with **six** domains. Domains **5** and **6** are strikingly similar to the helices formed by **U2/U6** and **U2/branch site**, respectively, in pre-mRNA splicing. It is thought that pre-mRNA splicing **evolved** from Group II splicing, and that spliceosome assembly creates an RNA catalytic center similar to that of Group II introns.

Different types of splicing

Some pre-mRNAs use **alternative** splicing to generate mRNAs with different coding capacities. Alternative splicing proceeds through the identical mechanism as 'constitutive' splicing, but different **splice sites** are chosen (the splice sites still obey the GT-AG rule). The *Drosophila transformer* gene, for example, uses a constant 5' splice site and one of two alternative 3' splice sites. In males, an upstream 3' splice site is used so that the next exon contains a **stop** codon, and Tra protein **is not** produced. In females, a 3' splice site is chosen that is **downstream** of the stop codon, so the Tra protein **is** produced.

While most splicing occurs within one RNA molecule, nematodes and trypanosomes can use *trans*-splicing. This splicing occurs when the **SL** RNA is spliced on to cellular mRNAs. The SL RNA has many features of a **snRNP** particle including binding to the same **Sm** proteins bound by the U-series splicing snRNPs. In trypanosomes, the SL RNA carries out the functions of the **U1** snRNP, which along with U5, is absent from these cells. *Trans*-splicing does use the same **U2**, **U4**, and **U6** snRNPs as *cis* splicing, indicating that the catalytic mechanism is the same in *cis*- and *trans*-splicing.

In contrast, tRNA introns are removed by a completely different mechanism using protein **endonuclease** and **ligase** enzymes. tRNA introns do not have **consensus** sequences at their splice sites. Rather, the intron is recognized by the **secondary** structure of the pre-tRNA.

3' end formation

RNA polymerase II transcripts are posttranscriptionally **processed** on their 3' ends. This requires the conserved sequence **AAUAAA** in the 3' untranslated region of the transcript. This sequence is bound by the factor **CPSF**, which then signals other **cleavage** factors to bind downstream. When a complete cleavage complex is assembled, the transcript is cleaved **downstream** of the AAUAAA. Subsequently, a **poly(A)** tail is added by the enzyme **poly(A) polymerase**. A few RNA polymerase II transcripts, such as the histone H3 mRNA, are not polyadenylated. The mature 3' end of these RNAs requires the **U7 RNA**, which base pairs with the histone mRNA to direct cleavage.

Chapter 31: Catalytic RNA

Answers to multiple-choice questions

1 (a,b,f); 2 (a,b,c,d); 3 (a,d,e); 4 (a,c,d,e); 5 (a,b,c,e); 6 (a,b,d); 7 (a,b,d); 8 (a,d,e).

Answers to concept questions

Question 1

The second step is identical in Group I and Group II splicing. It is initiated when the 3' OH group of the 5' exon attacks the 3' splice site. It results in joining of the exons and release of the intron. Note that the Group I and Group II introns are released in different forms: the Group I intron is linear with the exogenous G added to its 5' end while the Group II intron is in the lariat form.

Question 2

Group I introns, Group II introns, RNAase P, hammerhead ribozyme.

Question 3

RNA polymerase, endonuclease, phosphatase, ligase. The ability to convert the Group I intron into these enzymes suggests that it binds the sugar–phosphate backbone of RNA and can catalyze several different forward and reverse reactions on it. For example, ligation is the reverse of a cleavage reaction.

Question 4

Reverse transcriptase, endonuclease, maturase. These proteins create a DNA copy of the intron and make a double-strand break at a new site in the chromosome for the intron to be inserted.

Question 5

The hammerhead ribozyme binds a Mg^{2+} ion at the active site. This Mg^{2+} ion participates directly in the cleavage reaction by extracting a proton and attacking the cleavage site.

Question 6

Guide RNAs have a 5' region complementary to a pre-edited region of the mRNA, followed by a region complementary to the edited mRNA, with a poly(U) tail on the 3' end. Guide RNAs serve as templates that direct the insertion and deletion of U residues to create the edited mRNA.

Answers to problems

Problem 1

If you haven't thought of a brilliant experiment, first use this hint and try again: your experiment should use radioactive G.

Need another hint? Isolate total RNA from the mold under non-splicing conditions (no Mg^{2+}). Keep trying!

If you mixed this total RNA with the radioactive G and some Mg^{2+}, what would happen to the G if no Group I introns were present in the RNA preparation? Where would the G end up if a Group I intron were present? How could you tell the difference between free radioactive G (i.e., the starting material) and radioactive G that had been incorporated into a larger molecule?

Problem 2

What sequences are brought together if P1 pairs with the region near the 3' splice site?

Problem 3

Would you expect that a nucleotide involved in G-binding would be important for Group I splicing? Would it be present in all Group I introns? Would mutations in this nucleotide abolish G-binding and splicing? Michel and coworkers identified several nucleotides that fitted these assumptions, and then carefully studied mutations in the candidate nucleotides to discover which ones are directly involved in G-binding. They found that a specific G nucleotide, which is part of a G•C pair in the P7 helix, is primarily responsible for binding exogenous G. They went on to demonstrate that the exogenous G is bound by a triple strand interaction with the G of the P7 G•C pair.

Problem 4

First, the target RNA needs to contain the sequence 5'-GUN-3'. It should not be in a region of the RNA that takes part in any other RNA structure, such as a stem-loop, as that structure might preclude base pairing with the enzyme strand. The enzyme strand should pair with the target RNA, but is must not pair with any other cellular RNA, or it might cause cleavage of inappropriate RNAs. How would you know if your enzyme strand paired with any other RNA in the cell? Delivering an RNA drug to target cells is currently very challenging. RNAs could be delivered into cells by adding a gene encoding the enzyme strand, introducing this gene into cells, and allowing cellular RNA polymerase to make the RNA. Alternatively, the enzyme strand could be synthesized chemically and introduced into cells.

Problem 5

Did you account for both U insertion and deletion?

Answers to summary worksheet

Group I splicing

Group I splicing takes place in **two** steps. First, the **3' OH** group of an **exogenous guanosine** residue attacks the **5'** side of the first nucleotide of the intron. As a result, the exogenous **guanosine** residue is **added** to the intron and the first **exon** is liberated. Exon I now has a free **OH** group on its **3'** end. In the second step, this group attacks the **5'** side of the **first** nucleotide of the downstream exon. This reaction breaks the bond between the **intron** and the downstream exon and forms a new bond between the two **exons**.

All Group I introns adopt a common **secondary** structure with **nine** base-paired helices named **P1–P9**. The RNA sequence of most helices can **vary** as long as there are **compensatory** changes in the opposite strand. However, in some places, the **primary sequence** is conserved, indicating that these nucleotides are **essential** for Group I intron function. The conserved sequences **P** and **Q** base pair to form the **P4** stem while sequences **R** and **S** pair to form the **P7** stem. Along with the P3 and P6 helices, P4 and P7 form the **catalytic** core of the group I intron.

Some introns contain **open reading frames**. These encode proteins that allow the intron to **move** to new sites in the genome. These include an **endonuclease**, which makes a double-stranded break in the target DNA, and a **reverse transcriptase**, which copies the intron into DNA.

Ribozymes

In addition to splicing, RNA molecules can **catalyze** other reactions. Like proteins, RNAs adopt complex secondary and tertiary structures that form **binding** pockets for **metal** ions, cofactors and substrates. However, ribozymes generally have a **lower** catalytic rate than protein enzymes.

Other catalytic RNAs function as site-specific **endonucleases**. **RNAase P** is an enzyme that acts in tRNA processing. Like Group I and Group II introns, it contains both **protein** and **RNA** components *in vivo*. However, *in vitro*, the **RNA** alone is active. The **hammerhead** is an RNA motif that cleaves multimeric RNA genomes during replication of **viroids** and

virusoids. This forms an RNA structure that positions a Mg^{2+} ion at the catalytic site; this ion participates in catalysis by attacking the cleavage site.

RNA editing

RNA editing is an unusual set of reactions that alter the sequence of a mRNA **after** transcription. This reaction **is** highly specific. For example, in mammalian intestinal cells, a single **C** residue in the apo-lipoprotein B mRNA is changed to a **U** residue, creating a **stop** codon. This occurs by **deamination** of the C residue.

Editing in trypanosome **mitochondria** is much more extensive, involving **addition** and **deletion** of multiple U residues. This editing is directed by **guide** RNAs, which have three distinct regions. On the 5' end, the guide RNA is complementary to a **pre-edited** region of the mRNA. This sequence anneals to the pre-edited mRNA. The central region of the guide RNA is complementary to an **edited** region of the mRNA. This region directs the addition and deletion of U residues to create the **edited** sequence. At the 3' end, guide RNAs have a series of **U** residues. New evidence suggests that editing does not proceed by transesterification, but relies instead on protein enzymes to cleave the pre-edited mRNA, add or remove U residues, and ligate the pieces back together.

Chapter 32: Rearrangement of DNA

Answers to multiple-choice questions

1 (a,b,d); 2 (a,b); 3 (a,c,d); 4 (a,b,c,d,f); 5 (a,b); 6 (a,b,e); 7 (b,d); 8 (f); 9 (c); 10 (a,c,d,e).

Answers to concept questions

Question 1

The mating type system uses DNA rearrangement to move copies of mating type genes from the silent loci (*HMR* and *HML*) to the expressed locus (*MAT*). The silent loci are kept transcriptionally inactive by the chromatin structure at *HML* and *HMR*. The E and I silencer regions are insulators that create an inactive chromosomal domain. PRTF is an example of a transcription factor whose activity is changed by protein–protein interaction. PRTF can act alone to active **a**-specific genes. In combination with the $\alpha 1$ protein, it activates α-specific genes. When $\alpha 2$ is present, PRTF acts to repress **a**-specific genes. Cell type-specific gene expression is exemplified by the *HO* gene, which has a complex promoter that is active in haploid cells but not in diploid cells. The *HO* promoter is also regulated by the cell cycle, being active only at the end of phase G_1. Finally, the mating system uses a signal transduction cascade to control gene expression. Binding of mating pheromone to cell surface receptors activates a series of protein kinases that transmit the signal into the nucleus and alter gene expression (this type of pathway is discussed more thoroughly in Chapter 35).

Question 2

VSG proteins have a large N-terminal variable domain and a short C-terminal homology region.

Question 3

Surface antigen variation and mating type switching are alike in that each system has alleles at multiple loci in the genome, but only one locus is expressed at any time. In both systems, alleles can become active when they are copied into the expressed locus. However, there are more silent surface antigen loci (greater than one hundred) than silent mating type loci (two). Another difference is that the expressed locus for surface antigen genes is located at the telomere, while the expressed *MAT* locus is nearer the centromere. Finally, the expressed locus for surface antigen genes can change; this does not happen for mating type genes.

Question 4

Transfer of the T-DNA from bacterium to plant cells requires genes from both the bacterial genome and the plasmid. From the plasmid, the *VirA–E* and *VirG* genes are needed. Bacterial genes *chvA, chvB* and *pscA*, which encode the cell wall polysaccharides, are also required for transfer.

Question 5

Methotrexate is a drug that blocks folate metabolism. Cells can become resistant to this drug by increasing the copy number of (amplifying) the DHFR gene. In stably amplified cells, the genes are amplified at their normal chromosomal location. In unstable lines, the amplified genes form extrachromosomal elements called double minute chromosomes. These will be lost if methotrexate is removed.

Question 6

In transfection, DNA is injected into the nucleus of an oocyte or fertilized egg, which is then implanted into a pseudopregnant mouse. Some of the cells of the offspring may have incorporated the foreign DNA. Alternatively, DNA can be introduced into embryonic stem (ES) cells. These cells are at a very early stage of development and still retain the potential to develop into any tissue. The injected ES cells are mixed with a recipient blastocyst, which can then develop into a mouse.

Answers to problems

Problem 1

Are the left and right repeats the same after T-DNA integration? What does this say about the direction of copying?

Problem 2
What other prokaryotic regulatory processes are similar to *virA–virG*?

Problem 3
What happens when a yeast cell expressed both **a**1 and α2? The cell behaves like a diploid even though it has only a haploid DNA content.

Problem 4
What does HO do to the *MAT* locus? What is the consequence if this is not repaired?

Problem 5
This is a difficult but very important problem to solve, as trypanosome infections are a major health problem in many parts of the world.

Answers to summary worksheet

This chapter describes several ways in which cells rearrange DNA to alter gene expression. Yeast cells rearrange DNA to change their **mating type**. Trypanosomes can alter their surface coat by expressing a different **variable surface glycoprotein** gene. This is also accomplished by DNA rearrangement, and allows the parasite to escape the host **immune** response. Yeast mating type switching and trypanosome surface antigen variation are similar in several ways. Both systems have multiple **unexpressed** alleles of the rearrangeable gene but just one allele that is **expressed**. The expressed allele is located in one particular **locus**. The allele present in the expressed locus can be changed when one of the **unexpressed** alleles is **copied** into the expressed locus. This process requires a **double-strand break** in the chromosome at the expressed locus. In these DNA rearrangements, the donor unexpressed allele **is** retained at its original location.

Another type of DNA rearrangement occurs in the transfer of **T-DNA** from the bacteria *Agrobacterium* to plant cells. The T DNA causes growth of a plant **tumor** that secretes **opines**, which the bacterium uses for **food**. The **T-DNA** is copied from the **Ti plasmid** in the bacterium and transferred to the plant cell, where it **integrates** into the plant genome. This rearrangement does not involve a double-strand break; instead, a **single** strand of the T-DNA is excised by making a **nick** at each end.

DNA rearrangement occurs in mammalian cells when they are exposed to the drug **methotrexate**. Cells respond to this drug by **amplifying** the DHFR gene. In **stable** amplification, the new copies of the DHFR gene are located at their **normal chromosomal** position. **Unstable** amplification occurs when the new copies of the DHFR gene form extrachromosomal arrays called **double-minute chromosomes**.

DNA rearrangement can be exploited to introduce exogenous DNA into eukaryotic cells. New genes can be introduced into plant cells by cloning them into **T-DNA** and allowing *Agrobacterium* to transfer them into the plant genome. Several techniques can be used to introduce DNA into mammalian cells in order to make **transgenic** animals.

Chapter 33: Immune diversity

Answers to multiple-choice questions

1 (a,b,d,e); 2 (b,d,e); 3 (b); 4 (a,b,e); 5 (c); 6 (a,b,d); 7 (a,b,d,e).

Answers to concept questions

Question 1

Tolerance results from clonal deletion of B and T cells that recognize proteins found on the body's own cells. These cells are removed early during development.

Question 2

Heavy chain genes are assembled in several steps. First, a D region is joined to a J region. Next, a V region is joined upstream of the D region.

Question 3

Diversity is generated primarily through somatic recombination, which assembles millions of different antibody genes by combining V, J and constant region genes (heavy chains also use D regions for additional diversity). Addition diversity results from mutation of the immunoglobulin genes in active B cells.

Question 4

Killer T cells do not recognize their antigens directly: they must be 'presented' by another cell. The presenting cell digests the antigen into fragments, which are bound by the cell's major histocompatability complex (MHC) protein. The MHC protein with its bound antigen fragment then can interact with the T-cell receptor.

Question 5

The J segment on the 5' side of the J segment used in rearrangement is deleted. Any J segments on the 3' side are retained in the chromosome, but are removed by pre-mRNA splicing (they are part of the intron between the J and C segments).

Answers to problems

Problem 1

Recall that for each antigen, both immunoglobulin and T-cell receptor genes are needed. The human genome is estimated to contain approximately 250,000 genes, far short of the more than two million that would be required if these genes were encoded in their intact forms. This illustrates the remarkable efficiency of generating immunoglobulins and T-cell receptors by DNA rearrangement.

Problem 2

(c) Recall that heavy chains are composed of V, D and J segments. If the V region is followed by heptamer–12 nucleotides–nonamer, what organization must be found in front of the D segment? What organization must follow the D segment? What will be found preceding the J segment? What prevents a V segment being joined directly to a J segment?

Problem 3

(b) This seems like a reaction catalyzed by a DNA-dependent DNA polymerase. In what direction do all polymerases go? What direction is indicated here?

Problem 4

Trypanosomes 'win' because they have a much shorter cell division cycle. When a trypanosome infects a mammalian host, it duplicates in the bloodstream with a rapid doubling time. Soon after the infection begins, the resting B cell that recognizes the VSG of the infecting trypanosome is activated and begins to expand. However, mammalian cells divide much more slowly than trypanosomes. By the time the B cell has expanded enough to kill all the trypanosomes, some individual trypanosomes will be expressing a different VSG that is not recognized by the expanded B cell. These will start a new cycle of infection, growing until the immune system catches up and giving rise to new variants that will escape to start yet another cycle.

Answers to summary worksheet

The immune response relies on B and T **lymphocytes**, which produce proteins that specifically bind to foreign antigens. B cells make proteins called **immunoglobulins**. These proteins are composed of **two** heavy chains and **two** light chains, which have both **variable** and **constant** regions. Genes encoding intact immunoglobulin proteins **are not** found in the germline; rather, they are **assembled** from gene **segments**. By using different gene segments in different combinations, the immune system generates an enormous **diversity** of immunoglobulin proteins without requiring an enormous number of genes. Light chain genes are assembled from three segments known as **V** , **J** and **C**. Heavy chain genes contain an additional **D** segment, located between the **V** and the **J** segments.

The rearrangement of immunoglobulin genes is governed by the **heptamer** and **nonamer** consensus sequences found at the ends of all gene segments that can recombine. During rearrangement, the DNA is cut in the **heptamer** consensus sequence and joined to another **heptamer**. Additional nucleotides called **N** and **P**, which are not encoded in the genome, can be added at the junction to cause further alterations in the coding region.

Diversity of immunoglobulins results from several sources: different combinations of **V** and **J** gene segments (in addition to these, heavy chains also include **D** segments), different nucleotides at each segment **junction**, different combinations of **light** and **heavy** chains, and **somatic** mutation of the assembled immunoglobulin gene during the life of the B cell.

Not all immunoglobulin gene rearrangements are productive; some can generate genes that are incomplete or that change reading **frame**. Once a productive rearrangement has occurred, further rearrangements **are** prohibited on both copies of that locus (one on each copy of the chromosome). This is known as **allelic exclusion**, and ensures that each B cell makes only **one** immunoglobulin molecule.

Although the V-D-J junction of heavy chain genes does not change after a productive rearrangement has occurred, the **constant** region can change through the process of **class switching**. The different classes of C regions impart different **effector** functions, such as membrane anchoring or complement activation, to the immunoglobulin molecule without changing its antigen specificity.

T-cell receptors are similar to immunoglobulin molecules except that they contain **two** chains. T-cell receptor genes are assembled from gene segments in a process that **is** similar to immunoglobulin gene rearrangement. The T-cell receptor only recognizes its antigen when the antigen is bound by **MHC** protein. Other proteins encoded at the MHC locus include the **complement** proteins, which **lyse** cells bearing foreign antigens.

Chapter 34: Protein trafficking

Answers to multiple-choice questions

1 (a,c,d); 2 (a,b,d); 3 (a,c,d); 4 (a,b,c,d); 5 (b,d).

Answers to concept questions

Question 1

Secreted proteins fold into their proper tertiary structures in the ER. Recall from Chapter 10 that ribosomes synthesizing secreted proteins are brought to the ER by the SRP system (Figure 10.19). The nascent protein is passed through the ER membrane immediately after it is synthesized. Each domain folds sequentially. Proper folding requires, in some cases, the addition of oligosaccharides. The ER also contains accessory proteins that help secreted proteins to fold.

Question 2

(i) The protein is synthesized by ER-bound ribosomes.

(ii) The protein is folded in the ER.

(iii) N-linked glycosylation begins in the ER.

(iv) The protein is transported to the Golgi.

(v) N-linked glycosylation is completed in the Golgi.

(vi) O-linked glycosylation begins and is finished in the Golgi.

(vii) The oligosaccharides may be trimmed in the Golgi.

Question 3

The ARF protein initiates vesicle budding. It is myristoylated (modified by the addition of a lipid called myristic acid), which allows ARF to anchor in the membrane. ARF is activated by binding GTP, which allows it to bind coatomer. Additional coatomer assembles and the vesicle is invaginated.

Question 4

Vesicle fusion begins when the coatomer dissociates. Fusion requires the association of the soluble proteins NSF and SNAP with a membrane-bound receptor called SNARE. After ATP hydrolysis by SNAP, the vesicle fuses with the target membrane and NSF and SNAP are released.

Question 5

1. The receptor can be recycled to the cell surface in a coated vesicle while the ligand is destroyed in a lysosome. The LDL receptor is recycled through this type of receptor-mediated endocytosis.

2. Both the receptor and the ligand can be recycled, as is the case for transferrin and the transferrin receptor.

3. Both the receptor and the ligand can be degraded. An example of this type is provided by epidermal growth factor and its receptor.

4. The receptor–ligand complex can be shuttled through the cell and released by exocytosis on the other side of the cell. The immunoglobulin receptor uses this transcytosis mechanism to transport immunoglobulin across epithelial cells.

Answers to summary worksheet

Secreted proteins undergo posttranslational modification by the addition of **oligosaccharides**. Proteins are modified as they travel through the **endoplasmic reticulum** and the **Golgi**. When moving to a new location (either another organelle or to the cell membrane), proteins are carried in membranous **vesicles** that are surrounded by protein **coats**. Most of these vesicles contain the protein **clathrin**, which is important for the structure and function of the vesicle. These **coats** also contain specific proteins that direct the vesicle to its intended **target** membrane. **Receptors** are cell surface proteins that bind specific **ligands** and bring them into the cell in a process known as **receptor mediated endocytosis**.

Chapter 35: Signal transduction

Answers to multiple-choice questions

1 (b,c,d); 2 (c); 3 (a,b,c,d,e); 4 (a,b,c,e); 5 (b,c,d,e); 6 (a,c,d).

Answers to concept questions

Question 1

Carrier proteins bind an ion on one side of the membrane, change their conformation, then release the ion on the other side of the membrane. Thus, they carry the ion directly. Channel proteins form a hydrophilic pore in the membrane through which ions can pass. Both types of ion transporter are specific for certain ions and both can be regulated. In addition, they both need to have hydrophilic surfaces to associate with their ions. Both types of transporter use hydrophobic regions to shield these hydrophilic surfaces from the hydrophobic membrane.

Question 2

Trimeric G proteins are coupled to the cytoplasmic face of various receptors and are inactive. When a receptor binds its ligand, the G protein α subunit changes conformation and exchanges GDP for GTP. The α subunit then disassociates from the receptor and from the βγ subunits and activates an effector such as adenylate cyclase. The net effect is to increase the intracellular concentration of the second messenger cAMP.

Question 3

Receptor protein kinases are found in the cell membrane. They bind small peptides such as growth factors, which are secreted by other cells. They activate a variety of effector pathways that lead to the production of second messengers.

Question 4

Autophosphorylation stimulates kinase activity and causes the cytoplasmic domain to bind its target protein.

Question 5

PDGF, PDGF receptor, Grb2, SOS, Ras, GDP, GTP, Raf, MEK, MAP kinase, Elk-1.

Question 6

SOS activates Ras by stimulating it to exchange GDP for GTP. GAP inactivates Ras by stimulating GTP hydrolysis. GDI inactivates Ras by preventing GDP release.

Question 7

MAP kinases can transduce a signal into the nucleus by translocation of the kinase itself, by phosphorylating a factor that is translocated, or by phosphorylating an inhibitor, which allows a factor to be translocated.

Question 8

Trimeric G proteins such as Gs become active when the β-adrenergic receptor binds its ligand. The α subunit exchanges GDP for GTP, dissociates from the βγ subunits, and activates the effector. Monomeric G proteins such as Ras are activated when a receptor binds its ligand and is phosphorylated. This activates Grb2, which activates SOS, which in turn activates Ras by stimulating it to exchange GDP for GTP.

Question 9

The Ste5 protein is thought to be a scaffold to which a MEK kinase (Ste11), a MEK (Ste7) and a MAP kinase (Fus3) all bind. Thus, Ste5 interacts with the pathway at several levels.

Question 10

JAK kinases are activated when cytokine receptors bind their ligands and dimerize. The JAK kinase phosphorylates STATs, which are transcription factors. Phosphorylation causes STATs to dimerize and translocate to the nucleus, where they affect transcription. The JAK kinase pathway has fewer steps than the Ras pathway. Although the JAK pathway is simpler, it does not allow for as much signal amplification as the Ras pathway does.

Answers to summary worksheet: Chapter 35

Cells need to respond to external **signals**, either from their environment or from other cells. Many types of signaling molecules do not enter the cell. Rather, they interact with specific **receptors** found on the cell surface. These **receptors** activate a variety of **signal transduction** pathways that ultimately transmit the signal to the **nucleus**, where gene expression is altered.

One common pathway uses trimeric **G proteins**, so named because they bind **GTP**. These proteins interact with receptors when the receptors are in their **unbound** state but **dissociate** when the receptor binds its **ligand**. After dissociation, the α subunit acts on its target to stimulate the production of a **second** messenger such as **cAMP**.

Another pathway uses tyrosine kinase receptors, which **dimerize** upon ligand binding. This activates a **kinase** domain present in the **cytoplasmic** portion of the receptor. Some receptors activate the ras pathway. Ras is a **monomeric** G protein and **does not** bind directly to the receptor. It is activated when the **SOS** protein causes it to exchange **GDP** for **GTP**. Ras activates the **Raf** kinase, which in turn can activate a **MEK** pathway. The end result of these pathways is alteration in the pattern of **transcription**.

Chapter 36: Cell cycle and growth regulation

Answers to multiple-choice questions

1 (a,b,c,d); 2 (a,c,d); 3 (a,b,c); 4 (b); 5 (a,b,d).

Answers to concept questions

Question 1

The START checkpoint occurs in G_1 and commits the cell to S phase. To pass START, the cell must be over a certain mass. At the end of G_2, DNA integrity is checked to ensure that all the DNA has been replicated and that none of it is damaged. If a cell failed these checks and were allowed to proceed to mitosis, the daughter cells would probably not be viable.

Question 2

The p34 kinase becomes active when two phosphorylated amino acids in the ATP-binding domain (threonine-14 and tyrosine-15) are dephosphorylated. Threonine-161 must be phosphorylated for activity. The kinase is inactivated when the cyclin is destroyed.

Question 3

At START, cdc2 is active because it is phosphorylated on threonine 161, not phosphorylated on tyrosine 15, and compexed with cig2. It is inactivated by phosphorylation on tyrosine 15 and dissociation from cig2. In M phase, cdc2 becomes active again because it is phosphorylated on threonine 161, not phosphorylated on tyrosine 15, and complexed with cdc13.

Question 4

In G_1, RB binds to the transcription factor E2F, preventing the expression of genes required for S phase. At START, RB is phosphorylated by a cyclin/cdk complex and releases E2F, which activates transcription of genes required for S phase.

Question 5

We know that the chromosome, centrosome and cytoplasmic cycles are controlled coordinately because mutation of genes required for START, such as *cdc28*, stops all cycles. However, mutations in other genes block progress of only a single cycle, demonstrating the independence of some parts of each cycle. For example, *cdc31* mutants block only the centrosome cycle, mutation of *cdc7* specifically blocks the chromosome cycle, and *cdc24* mutants fail in only the cytoplasmic cycle.

Answers to summary worksheet

The cell cycle is controlled at several **checkpoints**. **START** occurs in G_1 phase and is the point at which the cell becomes committed to **DNA**

replication. Another checkpoint occurs in G_2 phase before the cell becomes committed to **mitosis**.

Mitosis is initiated by the **M phase kinase**, which contains **two** subunits. One subunit, called **p34** (the *S. pombe* homolog is called **cdc2** and the *S. cerevisiae* homolog is known as **CDC28**), is a kinase that phosphorylates specific target proteins. This subunit is regulated by **phosphorylation**. To be active, two amino acids within an **ATP**-binding domain must be dephosphorylated and one other amino acid must be phosphorylated. The other subunit is a **cyclin** that also regulates the activity of the catalytic subunit. These **cyclins** accumulate throughout the cell cycle but are rapidly **degraded** during **M** phase. Without the **cyclin** subunit, the **catalytic** subunit is inactive.

The transition from G_1 to S phase is controlled by a kinase that contains the same **catalytic** subunit but a different **cyclin** subunit. Numerous other proteins, originally identified as **cdc** (which stands for **cell division cycle**) mutants, are required for the cell cycle. These proteins function to duplicate and partition the **chromosomes** and **centrosomes**. Upon entry into **M** phase, the cell is dramatically **reorganized**. This process includes **condensation** of chromatin, breakdown of the **nuclear envelope**, **endoplasmic reticulum** and **Golgi**, and reorganization of microtubules into the **spindle**.

Chapter 37: Oncogenes and cancer

Answers to multiple-choice questions

1 (a); 2 (a,b,c,d); 3 (a,b,d); 4 (a,c,e,f); 5 (b,c,d).

Answers to concept questions

Question 1

Viral *onc* genes may cause oncogenesis because they are expressed at higher levels than cellular *onc* genes. Alternatively, v-*onc* genes can carry mutations that alter the activity of the encoded protein.

Question 2

Wild-type retroviruses can cause oncogenic transformation when they insert into the host chromosome. The net result is the inappropriate expression of a c-*onc* gene. If the provirus inserts into the first intron of a c-*onc* gene (and inserts in the same orientation as the c-*onc* gene), c-*onc* will be transcribed by the strong promoter in the retroviral LTR. This increases the level of c-Onc protein and removes c-*onc* from its normal transcriptional control. If the provirus inserts in the opposite direction, it could activate a fortuitous promoter that transcribes c-onc. Even insertions downstream of c-*onc* can cause it to be expressed if the viral enhancer stimulates the normal v-*onc* promoter.

Question 3

1. Growth factors such as Sis can become oncogenic (but only to cells that express the receptor for that growth factor). Presumably, these mutant growth factors stimulate their receptors inappropriately.
2. Growth factor receptors like c-ErbB can become oncogenic if a mutation activates their kinase domain constitutively.
3. Intracellular protein kinases such as Src can become oncogenic by mutations that activate the kinase domain or inactivate the suppressor domain.
4. Signal transduction molecules like Ras can become oncogenic if they have a mutation that interferes with their normal regulation.
5. Cytoplasmic serine/threonine kinases, for example, Mos, are made oncogenic by mutations that activate their kinase domains.
6. Transcription factors like Myc can become oncogenic, presumably by causing expression of genes involved in cell division at inappropriate times.

Question 4

Some transcription factors can be oncogenic if their activity or rate of synthesis is increased, of if they escape their normal regulation. For example, Fos and Jun are leucine zipper proteins that make up the AP1 transcription factor (see Chapter 29). Each protein can become oncogenic, generally by truncations, deletions and point mutations. These mutants remove domains of the proteins that are required for their normal regulation. Another class of transcription factors become oncogenic when they lose their ability to activate transcription.

Question 5

Deletion of the thyroid hormone binding domain of ErbA renders it oncogenic. Without the hormone-binding domain, this protein cannot activate transcription. In fact, the oncogenic form inactivates transcription by forming inactive heterodimers. Oncogenesis is thought to result from the loss of tumor suppressor gene products that depend on ErbA for their transcription.

Question 6

p53 contains a transcription activation domain, a DNA-binding domain that recognizes a specific 10 base pair palindrome, an oligomerization domain, and a C-terminal domain that binds nonspecifically to damaged (single-stranded) DNA. Most oncogenic mutations map to the site-specific DNA-binding domain.

Answers to summary worksheet

Oncogenesis can result from the inappropriate expression of *onc* genes, which causes cells to divide when they should not. Retroviruses that acquire cellular genes (**c-*onc*** genes) can cause oncogenesis by **overexpression** of the wild-type gene. Other **transducing** viruses can be oncogenic by expressing a **mutant** form of the gene. Even retroviruses that do not carry a **v-*onc*** gene can be oncogenic if they **integrate** into the host chromosome near a **c-*onc*** gene and alter its **expression**. *Onc* genes can also be created by chromosome **translocations**.

Many *onc* genes encode proteins that function in **signal transduction** pathways. Proteins at any point in the pathway can become oncogenic including **growth** factors, **growth** factor **receptors**, G proteins (**Ras**, for example), intracellular **kinases** and other signaling molecules. These factors become oncogenic when they **mutate** to a form that **does not** respond to their normal control mechanisms. Ultimately, oncogenes cause inappropriate **transcription**. In fact, many **transcription** factors themselves can mutate to oncogenic forms.

The tumor suppressor **RB** functions by sequestering the transcription factor **E2F**, which activates genes required for **S** phase. The tumor suppressor **p53** can stop the cell cycle in G_1 (for example, if DNA is **damaged**). This protein can also trigger **apoptosis**.

Chapter 38: Gradients and cascades

Answers to multiple-choice questions

1 (a,b); 2 (b); 3 (a,b,d,f); 4 (a,b,d,e); 5 (a,b,c); 6 (a,b,c,d,f).

Answers to concept questions

Question 1

The anterior system is required for the formation of the head and thorax. It creates a gradient of *bicoid* mRNA. The posterior system is responsible for the development of abdominal structures. It acts to localize the morphogen Nanos. The terminal system directs the development of the very ends of the egg. It acts through the Torso transmembrane receptor.

Question 2

Hunchback is not expressed unless Bicoid is present above a certain threshold level. Thus, the Bicoid gradient acts like a binary switch for the expression of *hunchback*.

Question 3

Dorsal–ventral pattern is established by a signal that activates the ubiquitous membrane-spanning receptor Toll only on the ventral side of the egg.

Question 4

Mutations in gap genes cause deletion of a group of segments from the body plan. Pair-rule gene mutations cause deletion of parts of the pattern from alternative segments. Mutation of segment polarity genes replaces the posterior compartment of each segment with a mirror image of the corresponding anterior compartment.

Question 5

Loss-of-function alleles cause posterior segments to transform into anterior segments, while gain-of-function alleles (such as alleles that cause overexpression) cause anterior segments to have more posterior characteristics.

Question 6

Hox genes share sequence homology with the *ANT-C* and *BX-C* genes, and are found in similar clusters. The mouse genes, like their *Drosophila* counterparts, are arranged in the order of their expression pattern. These similarities argue that the *Hox* and *ANT-C/BX-C* genes are closely related through evolution. *Hox* genes are smaller than the *Drosophila* homeobox genes, and there are more of them, including some duplicated genes.

Answers to summary worksheet

Development of a morphologically complex organism from a fertilized egg requires the action of many sets of genes. This process has been studied extensively in *Drosophila melanogaster*. **Maternal** genes are expressed in the mother and packaged into the egg. These gene products establish the **anterior–posterior** and **dorsal–ventral** axes of the egg. For example, the *bicoid* mRNA is present in a concentration **gradient** in the egg. In places where the bicoid concentration is above a certain threshold, it activates the *hunchback* gene, leading to the development of **anterior** structures.

Segmentation genes are expressed after fertilization. These are found in three types known as **gap** genes, **pair-rule** genes, and **segment polarity** genes. They are expressed in patterns that divide the developing embryo into segments.

Finally, **homeotic** genes determine the ultimate development of each segment. These are encoded by **complex** loci such as *ANT-C* and *BX-C*. All these sets of genes work together in a combinatorial fashion to give each part of the embryo a specific developmental fate.